Biology of Earthworms

Biology of Earthworms

C. A. Edwards

J. R. Lofty

Rothamsted Experimental Station, Harpenden

London

CHAPMAN AND HALL LTD

First published 1972
by Chapman and Hall Ltd
11 New Fetter Lane, London EC4P 4EE

Filmset in Photon Imprint 11 on $12\frac{1}{2}$ pt. by
Richard Clay (The Chaucer Press), Ltd., Bungay, Suffolk
and printed in Great Britain by
Fletcher & Son Ltd., Norwich

SBN 412 11060 1

Distributed in the U.S.A. by
HARPER & ROW PUBLISHERS, INC.
BARNES & NOBLE IMPORT DIVISION

Contents

Foreword

by Dr J. E. Satchell
Merlewood Research Station,
Grange-over-Sands, Lancashire

As the classroom type of the Annelida, 'the earthworm' is familiar to every student of biology. Dissection manuals neatly package its anatomical features and numerous texts admirably relate earthworm form and function. Beyond these frontiers, student and teacher alike are met by a bewildering mass of publications ranging in content from the military value of earthworms as survival food, to the species stored by moles. The volume of this literature is quite exceptional for an invertebrate group which is important neither as pest nor food. It stems from the inherent zoological interest of the earthworm as a terrestrial form retaining many of the characteristics of its aquatic ancestors; from its convenience as an experimental animal for behaviourists and physiologists; and from its effects on soil fertility.

This last aspect had already been perceived two centuries ago by Gilbert White of Selborne who, in 1770, wrote, 'Earthworms, though in appearance a small and despicable link in the chain of Nature, yet, if lost, would make a lamentable chasm . . . worms seem to be the great promoters of vegetation, which would proceed but lamely without them, . . .' In more recent times, such ideas have been critically examined and extended at Rothamsted Experimental Station in a programme of earthworm research pioneered by Sir John Russell and continued by a number of workers during the last twenty-five years. It is particularly appropriate that the task of surveying the earthworm literature and summarising it at readable length should have been undertaken by the current representatives of this Rothamsted tradition.

An important part of this book which will be particularly valuable to soil zoologists, is the bibliography of between five and six hundred literature references. An analysis of their frequency distribution in time throws an interesting light on the development of earthworm studies. Divided into twenty-year periods they fall as follows: 1870–1889 6; 1890–1909 12; 1910–1929 68; 1930–1949 106; 1950–1969 361. Allowing for some selection in favour of more recent studies, it seems that, whereas around the turn of the century earthworm papers were published on average about one every other year, in the last two decades one was published on average about every three weeks. If publication continues at this exponential rate, a revision of this book in twenty years' time will involve the formidable task of reviewing another 640 papers.

In the face of this alarming publication explosion it is interesting to note how slowly concepts and attitudes change, even amongst contemporary ecologists. Charles Darwin's famous book, *The Formation of Vegetable Mould through the Action of Worms*, published in 1881, provides a good example. When Henry James described the requirements for a successful mid-Victorian genre painting as 'that it shall embody . . . some comfortable incident of the daily life of our period, suggestive more especially of its . . . familiar moralities', he might well have had in mind Darwin's analogies of earthworms as beneficent gardeners and industrious ploughmen. It is significant to find this anthropocentric approach epitomised almost a century later in this present book in the chapter title 'Earthworms as benefactors'. Indeed, as the authors show, some of our ideas about earthworms can be traced back to Aristotle.

Despite the voluminous literature of the intervening period, E. W. Russell, summarizing our knowledge of the importance of earthworms in agriculture, wrote in 1950, 'They may play an important role in the conversion of plant into humus . . ., but this has not yet been rigorously proved'. Twenty-two years later this is still substantially true for we now know that earthworm activity is primarily important through its effects on soil microflora and these effects are extremely difficult to measure outside the artificial conditions of laboratory cultures. Advances in this field await developments in the techniques of microbiology.

The prospective student of earthworm ecology should not be inhibited by the weight of past research for much remains to be done. Important land-use changes may be expected in Europe in the coming decades with the restoration of former industrial sites and the withdrawal of five million hectares from agricultural use under the Mansholt plan. It is in this marginal and restored land, rather than in the agricultural lowlands, that earthworm research is likely to be most relevant and fruitful for it is under the less intensively managed land on base rich sites that soil processes generally become dominated by earthworm activity. Outside the temperate zone and particularly in the tropics, much progress has been made in recent years in the basic taxonomy of indigenous earthworm species, but knowledge of their functional role as components of ecosystems is, at the best, fragmentary. Dr Edwards' and Mr Lofty's review of the current state of earthworm knowledge will prepare the way for another generation of research in these diverse and important fields.

J. E. Satchell

1. Morphology

1.1 Segmentation: external

Earthworms are divided externally into bands or segments along the length of the body by furrows or intersegmental grooves, which coincide with the positions of the septa dividing the body internally.

Plate 1 Prostomium of *A. caliginosa*

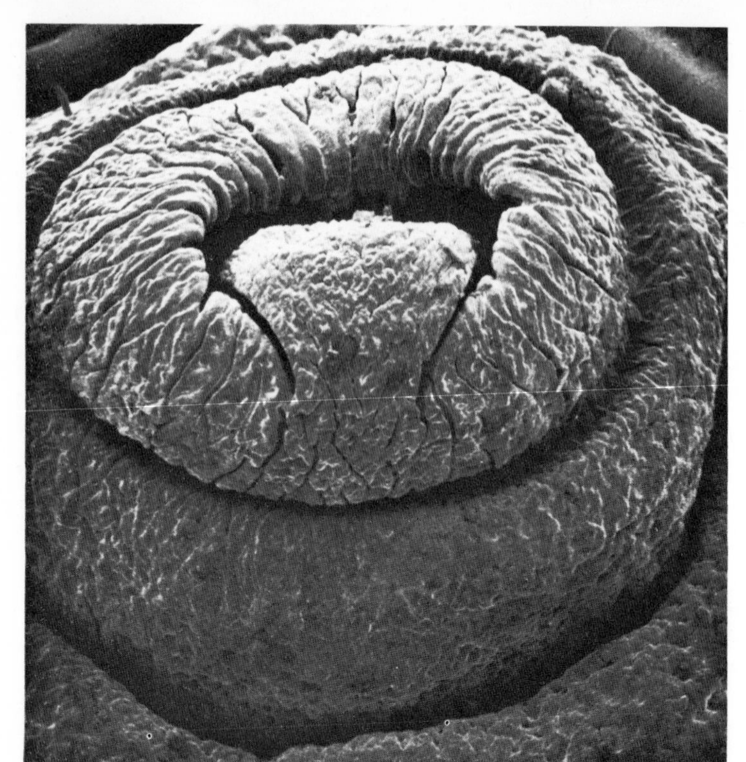
2(a)

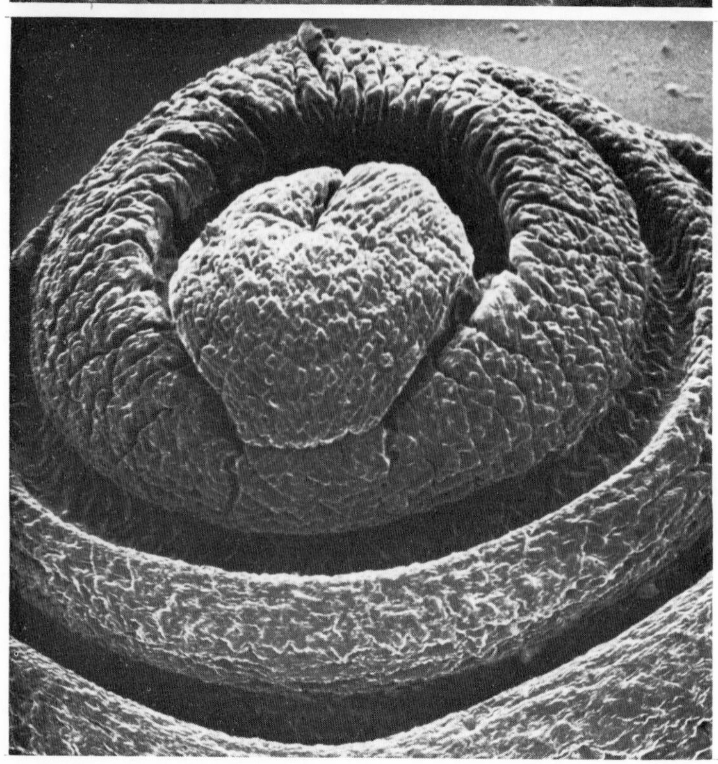
2(b)

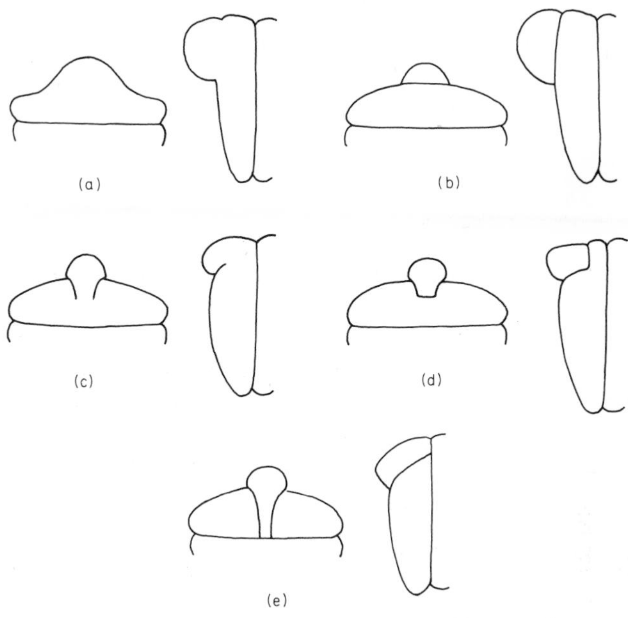

Fig. 1 Various forms of prostomium (cephalisation)
(*a*) zygolobous (*b*) prolobous (*c*) and (*d*) epilobous (*e*) tanylobous.

The segments vary in width, usually being widest in the anterior and clitellar regions. Segments are arbitrarily numbered from the front to the rear, and the grooves are designated by the numbers of the segments on either side, e.g. 3/4, 10/11, etc. Often, the external segments are subdivided by one or two secondary grooves, particularly in the anterior part, but these are superficial divisions which are not reflected in the internal anatomy. The mouth opens on the first segment, or peristomium, which bears on its dorsal surface the prostomium, a lobe overhanging the mouth (plates 1, 2a and b). The prostomium varies in size, and in some worms it may be so small that it cannot be distinguished. The way the peristomium and the prostomium are joined differs between species and is a useful systemic character. The connection is termed zygolobous, prolobous, epilobous or tanylobous, depending on the demarcation of the prostomium (Fig. 1). Some of the aquatic worms (Naididae and

---

Plate 2a Epilobous prostomium of *A. chlorotica*
Plate 2b Epilobous prostomium of *A. caliginosa*

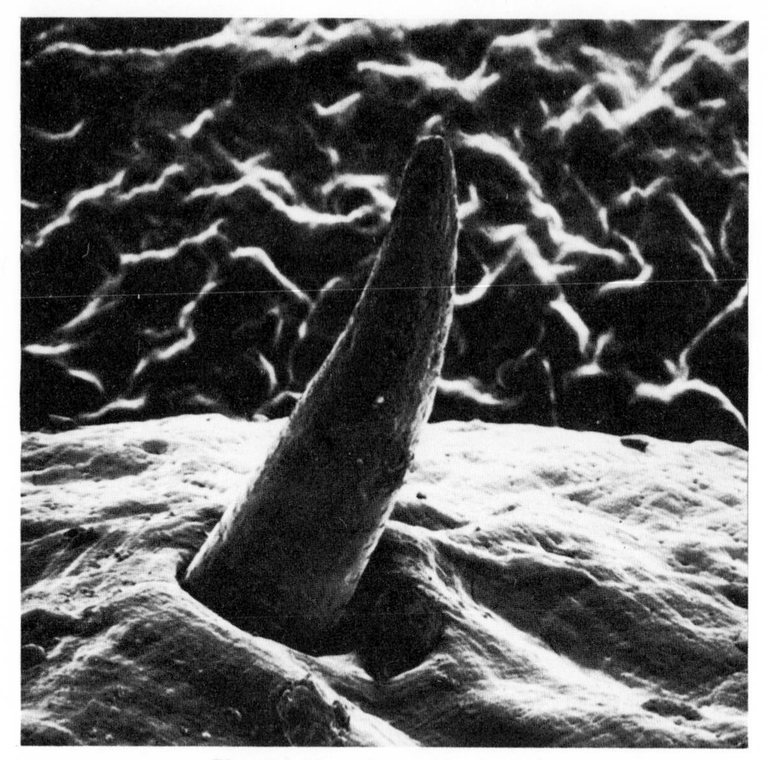

Plate 3a Normal seta of *L. terrestris*

Lumbriculidae) have the prostomium extended forward into a proboscis.

Sometimes there is unusual segmentation in the region just before the anus. *Pontoscolex* (Glossoscolecinae) has a swollen posterior end, with narrow segments; it has been suggested this is a tactile sensory region adapted for gripping the sides of the burrow to avoid being pulled from the ground.

1.2 Chaetotaxy

The setae, which are bristle-like structures borne in follicles on the exterior of the body wall, can be extended or retracted by means of protractor and retractor muscles which are attached to the base of the follicles and pass through the longitudinal muscle layer into the

Plate 3b Genital seta of *A. caliginosa*

circular muscle layer below. As the setae are used to grip the substrate, their principal function is locomotory. Different species of oligochaetes have setae of varying shapes – either rod, needle or hair-like (Plate 3a). Rarely, the distal end is forked. The shape of setae varies with their position, the commonest form being those of *Lumbricus*, which are sigmoid and about 1 mm long. Often setae are enlarged both at the anterior and posterior ends, as in *L. terrestris*; setae in the region of the genital pores (particularly the male pores) are sometimes modified in size and shape, and situated on raised papillae (Plate 3b). These genital setae, which can be up to 7 mm long in some species of Lumbricidae, are usually grooved along their length, and may have a hook-like process at the distal end. These setae assist the physiological processes which take place at

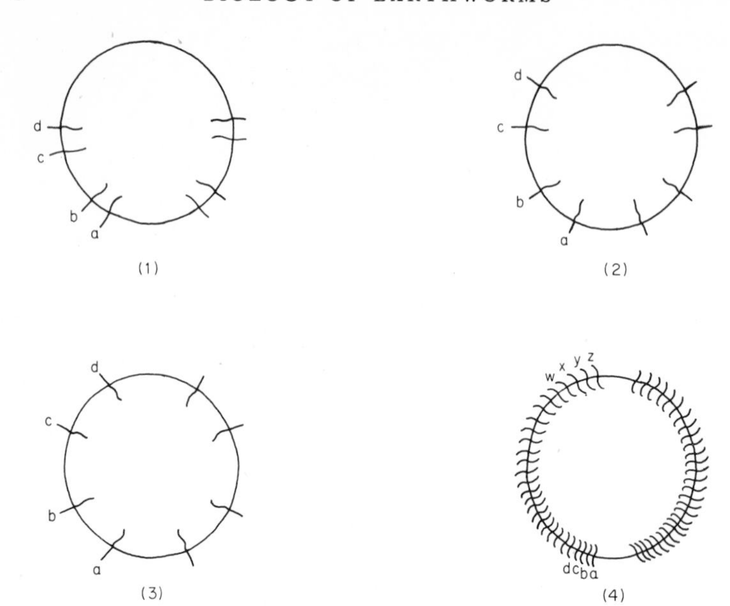

Fig. 2 Arrangement of setae in Oligochaeta
1, 2, 3, lumbricine arrangement 4, perichaetine arrangement
1. closely-paired 2. widely-paired 3. distant-paired.

copulation by providing physical stimuli to the partner, while other setae have been shown to assist in holding copulating earthworms together by gripping, clasping or penetrating the skin (Feldkamp, 1924).

The setae are arranged in a single ring around the periphery of each segment, their number and distribution being typically termed either lumbricine or perichaetine. The lumbricine arrangement, as in the Lumbricidae, is 8 per segment in ventral and latero-ventral pairs. If the distance between the setae in each pair is very small, they are termed 'closely paired', if wider apart, they are termed 'widely paired', or if they are very far apart so that the pairing is not obvious, they are termed 'distant'. The distance between each pair and between neighbouring pairs is constant for each species. The setae are designated by letters, *a*, *b*, *c* and *d* beginning with the most ventral setae on each side (Fig. 2). The distance between the setae is important as a systemic character and is usually expressed as the ratio of distances between setae. For example,

$aa : ab : bc : cd : dd = 16 : 4 : 14 : 3 : 64$ is the ratio in the post-clitellar region of *L. terrestris.* Alternatively, the distances may be expressed as an equation, $aa = 4ab$; or as $ab < bc > cd$. The distance between the two most dorsal setae on either side is usually compared with the circumference of the body (u) at that point, thus $dd = \frac{1}{2}u$. For many species of the Megascolecidae, particularly *Megascolex*, *Pheretima* and *Perionyx*, and certain others, the setal arrangement is termed perichaetine. Here the setae are arranged in a ring right round the segment, although usually with a large or smaller break in the mid-dorsal and mid-ventral regions. (Both types of setal arrangement are shown in Fig. 2.) In these worms there are more than eight setae per segment, often 50–100, sometimes more. Setal distributions intermediate between lumbricine and perichaetine may be 12, 16, 20 or 24 setae per segment, fairly distinctly arranged in 6, 8, 10 or 12 pairs, or the arrangement may be lumbricine in the anterior and appear perichaetine in the posterior part of the body. The setae are designated by the letters *a*, *b*, *c*, *d* . . . beginning with the most ventral one on each side, and *z*, *y*, *x*, *w* . . . beginning with the most dorsal one on each side, irrespective of how many there are in between.

1.3 Genital and other apertures

Earthworms are hermaphrodite and have both male and female genital openings to the exterior, as paired pores on the ventral or ventro-lateral side of the body (Plate 4). In lumbricid worms the male pores are situated ventro-laterally on the 15th, or occasionally on the 13th segment. Each pore lies in a slit-like depression, which in some species is bordered by raised lips or glandular papillae often extending on to the neighbouring segments. In other families, the male pores may be on quite different segments, and in some, particularly the Megascolecidae, the male pores are often associated with one or two pairs of prostatic pores; these are openings of the ducts of accessory reproductive bodies known as prostates, which are usually absent from lumbricids. Both the male and prostatic pores may be on raised papillae or ridges, or may open directly on to the surface. The male and prostatic pores are sometimes combined as one opening, but when separate they are usually joined by

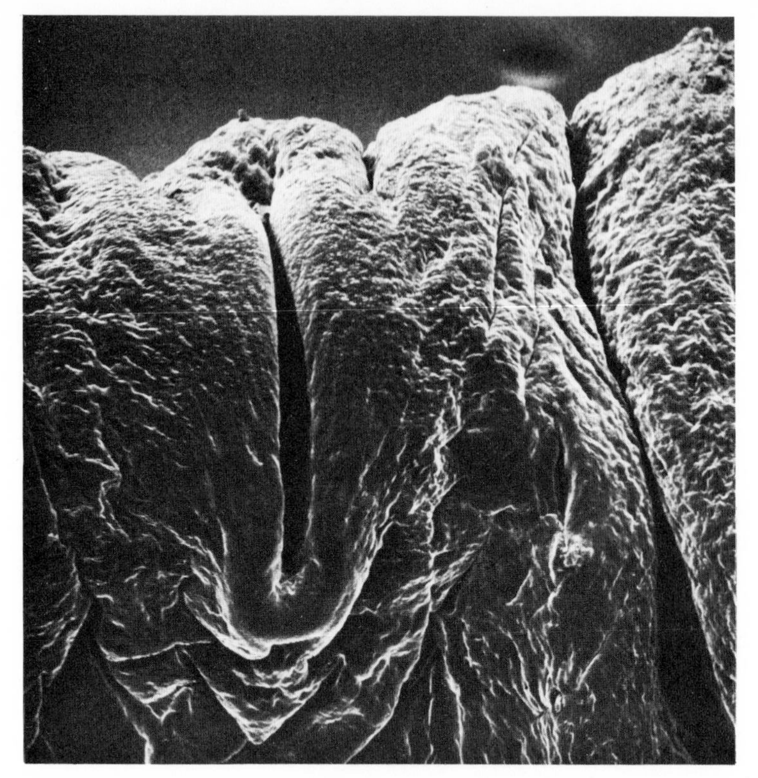

Plate 4 Male and female apertures of *A. caliginosa*

longitudinal seminal grooves, on either side of the ventral surface of the body.

Usually earthworms have two or more pairs of spermathecal pores, with a maximum of seven pairs in some species, but in others they are absent (*Bimastos tenuis*, *B. eiseni*). Worms from the Enchytraeidae and two or three other families, have only a single pair of pores. Spermathecae and their pores are not always paired. For instance, some megascolecid worms, particularly the genus *Pheretima*, have a single series, situated in the mid-ventral line. In the same genus are species with many spermathecae per segment, for example, *P. stelleri* has as many as thirty. Spermathecal pores are usually intersegmental and are most often in the ventral or latero-ventral position, but sometimes they are close to the mid-dorsal line.

The female pores are most commonly a single pair, situated either in an intersegmental groove or on a segment, their position often being diagnostic of a particular family. Thus, in the Enchytraeidae they are in groove 12/13, and in the Lumbricidae, Megascolecidae and Glossoscolecidae they are on segment 14. Sometimes the female pores are united into a single median pore.

The dorsal pores, which are small openings situated in the intersegmental grooves on the mid-dorsal line (Plate 5a), occur in most terrestrial oligochaetes, but not in aquatic and semi-aquatic species. These pores communicate with the body cavity and the coelomic fluid. There are usually no dorsal pores in the first few intersegmental grooves. The position of the first dorsal pore is used as a systemic character, at species level, although these pores are hard to distinguish in some worms. Some species have pores which are readily visible, and some, e.g. *Allolobophora rosea*, have pores surrounded by a dark pigmented zone.

The other openings in the body wall are the nephridiopores, which are very small and difficult to see. They are situated just posterior to the intersegmental grooves on the lateral aspect of the body, and usually extend in a single series along the body on either side. The nephridiopores are the external openings of the nephridia (the excretory organs of the earthworm) and, like the dorsal pores, are normally kept closed by sphincter muscles.

1.4 The clitellum and associated structures

The clitellum is the glandular portion of the epidermis associated with cocoon production. It is either a saddle-shaped or annular structure, the former being most usual among the Lumbricidae. It usually appears swollen, although sometimes it can only be differentiated externally by colour. In some megascolecids it appears only as a well-defined constriction (Stephenson, 1930). When swollen, as it usually is in mature lumbricids, the intersegmental grooves are often indistinct or entirely obscured, particularly on the dorsal surface. The clitellum is usually paler or darker than the rest of the body, or it may be a different colour.

The position of the clitellum and the number of segments over which it extends, differs considerably among oligochaetes. Lumbricidae have the clitellum on the anterior part of the body behind

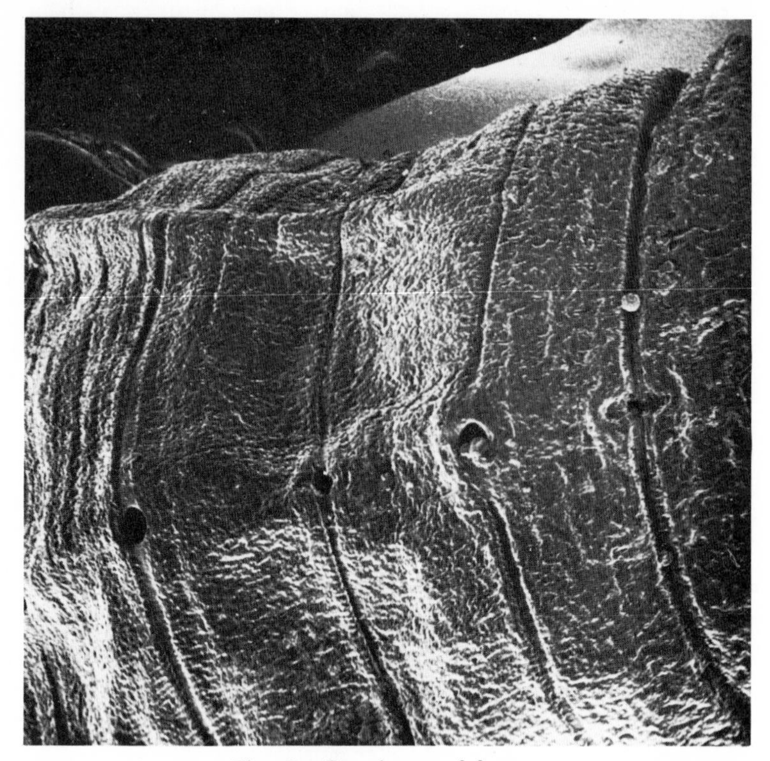

Plate 5a Dorsal pores of *A. rosea*

the genital pores, beginning on segments between 22 and 38, and extending over about four to ten segments posteriorly (Fig. 3). Megascolecidae have the clitellum further forward, beginning at or in front of segment 14, thus including the female pore, and posteriorly it may also include the male pore. Some aquatic or semi-aquatic worms, and also the Enchytraeidae, have a clitellum which is only a very temporary development during the period of cocoon formation, while even in the Lumbricidae it is often only conspicuous during the breeding season.

Most earthworms possess various markings at sexual maturity, in the form of tubercles, ridges and papillae on the anterior ventral surface, and these differ greatly in number and form in different species of oligochaetes. The tubercula pubertatis (Plate 5b) are glandular thickenings on the ventral surface, either on or near

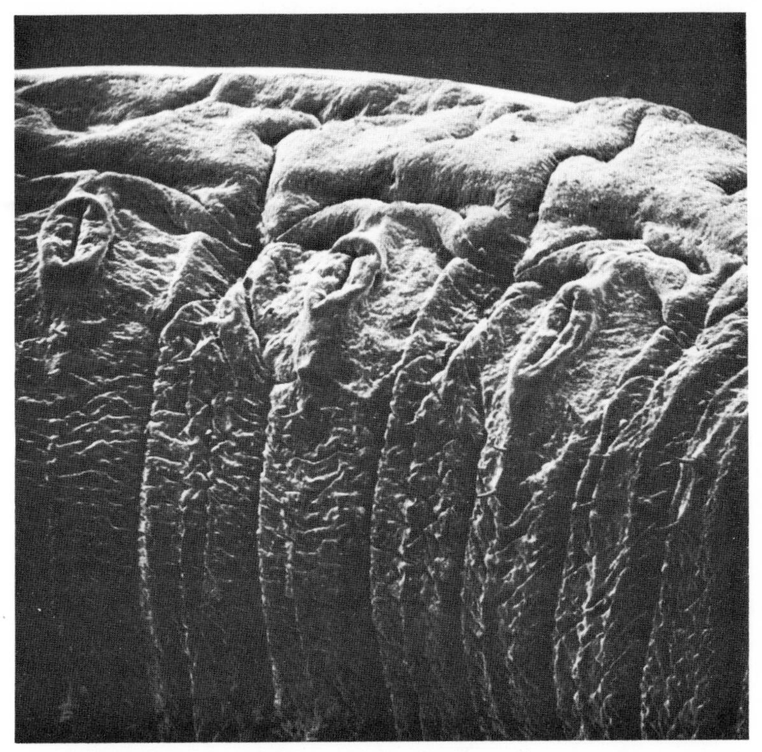

Plate 5b Part of clitellum and tubercula pubertatis of *A. chlorotica*

the clitellum. Lumbricidae have these in the form of a pair of more or less oval longitudinal ridges, which are sometimes partly divided by intersegmental grooves, or they separate papillae on either side of the ventral surface of the clitellum. The tubercula pubertatis usually extend over fewer segments than those occupied by the clitellum, except in some species of lumbricids where they extend beyond the clitellum. Those species without spermathacae, e.g. *B. eiseni*, usually have no tubercula pubertatis. It has been suggested that these organs assist in keeping copulating worms together (Benham, 1896). Worms of other families do not necessarily have the tubercula pubertatis situated on the clitellar segments, for example, in the Megascolecidae they are variously on either side or on both sides of the clitellum, or on the clitellar segments. They may be transverse ridges or small separate pads,

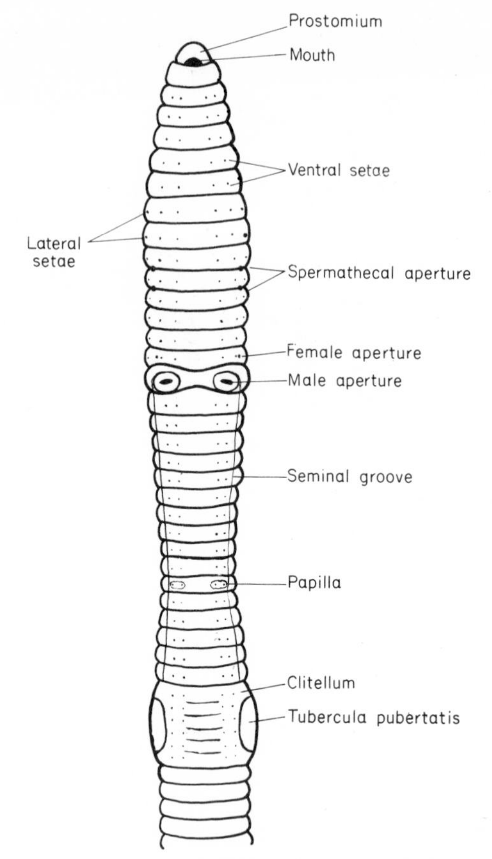

Fig. 3 Ventral view of the anterior region of *L. terrestris*. (*After Stephenson, 1930*)

symmetrically (sometimes assymmetrically) arranged about the mid-ventral line.

The position of the clitellum is used as a diagnostic character, particularly for lumbricids, because the position and number of segments occupied by the clitellum is, with small variations, constant for each species. Its position is defined by the number of the first and last segments occupied, the range of variations in position thus: 26, 27–31, 32.

1.5 Pigmentation

The colour of earthworms depends mostly on the presence or absence of pigment, which is either in the form of granules or in pigment cells in the subcuticular muscle layer. Pigmented worms, such as *L. terrestris*, are usually red, brown, a combination of these colours, or even greenish. Occasionally the colour is at least partly due to the presence of yellow coelomic fluid or green chloragogen cells near the surface. The ventral surface of these worms is usually much lighter in colour than the dorsal surface, although some deeply pigmented megascolecids are equally pigmented on both dorsal and ventral surfaces. The pigments, which are mainly porphyrins, probably originate as breakdown products of the chloragogen cells (Stephenson, 1930). Unusually, the pigment is not evenly distributed, but appears as dark segmental bands separated by lighter intersegmental zones, as in *Eisenia foetida*. Members of the Aeolosomatidae (small freshwater worms) have globules, coloured yellow, red, orange or green in the epidermis. Lightly pigmented or unpigmented worms often appear reddish or pink, due to the haemoglobin of the blood in the surface capillary vessels showing through the transparent body wall. If the body wall is opaque, unpigmented worms are whitish. In a number of species the cuticle is strongly irridescent and causes the worms to appear bluish or greenish. This is particularly noticeable in species of *Lumbricus* and *Dendrobaena*. The colour of pigmented worms when preserved in formalin is fairly stable, but the reds and pinks of unpigmented worms usually fade rapidly.

1.6 The body wall

The body wall consists of an outer cuticle, the epidermis, a layer of nervous tissue, circular and longitudinal muscle layers, and finally the peritoneum, which separates the body wall from the coelom.

The cuticle, which is a very thin (7 μm in *L. terrestris*) non-cellular layer, is colourless and transparent, consisting of two or more layers, each composed of interlacing collagenous fibres, with several homogenous non-fibrous layers beneath. It is perforated by many small pores which are larger and more irregularly distributed about the middle of each segment than in its anterior part, and are

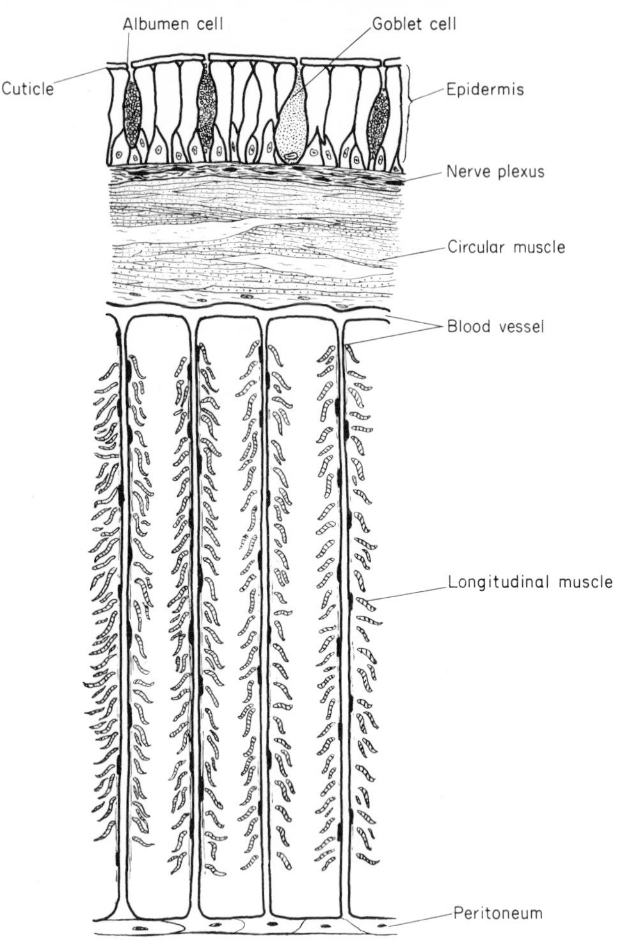

Fig. 4 Transverse section of a portion of the body wall. (*After Grove and Newell, 1962*)

absent from the posterior part of each segment. The cuticle is thinnest where it overlays the epithelial sense organs, and here it is perforated by many very small pores through which project fine hairs from the sensory cells.

The epidermis consists of a single layer of several different kinds of cells (Fig. 4). The supporting cells, which are columnar in shape, are the main structural cells of the epidermis, and have processes

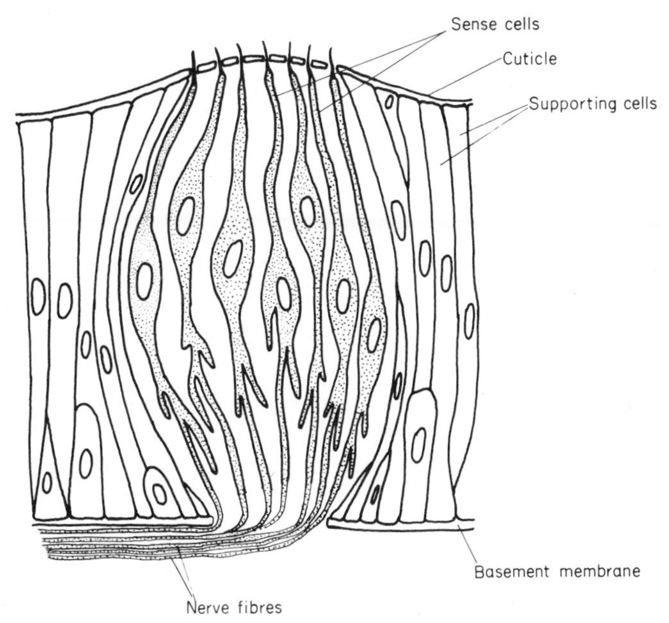

Fig. 5 Vertical section of an epidermal sense organ. (*After Grove and Newell, 1962*)

which extend into the muscle at their bases. These cells secrete material to form the cuticle. Short and round, or club-shaped, basal cells lie on the inner wall of the epidermis, and are also known as 'replacing cells', although it is doubtful whether they can replace the epithelial cells.

Two forms of glandular cells present are the mucous or goblet cells and albumen cells. Mucous cells secrete mucus over the surface of the cuticle to prevent desiccation and to facilitate movement through soil. The function of the albumen cells is not known.

Large numbers of sensory cells, grouped together to form sense organs which respond to tactile stimuli, are scattered throughout the epidermis (Figs. 5 and 6). These are more numerous on the ventral than on the dorsal surface. Photoreceptor cells, capable of distinguishing differences in light intensity, occur in the basal part of the epidermis (Fig. 6), and are most numerous on the prostomium and first segment and on the last segment. There are few or none in the epidermis of the middle segments. The epithelium of the buccal cavity bears groups of sensory cells which can be stimulated by chemical substances associated with taste. The prostomium contains receptors which can detect sucrose, glucose and also quinine (Laverack, 1959).

The epidermis is bounded at its inner surface by a basal membrane within which lie two muscle layers. The circular muscle layer consists of muscle fibres extending around the circumference of the body, except at the intersegmental positions. The muscle fibres are arranged in an irregular manner, aggregated into groups, each surrounded by a sheath of connective tissue (Fig. 4).

The longitudinal muscles, which are in a much thicker layer than the circular muscles, are continuous throughout the length of the body, and arranged in groups or blocks around the body. ***Lumbricus*** has nine such blocks of muscles, two dorsal, one ventral and three ventro-lateral on either side. They have ribbon-like fibres which are arranged into U-shaped bundles in each block, these are surrounded by sheaves of connective tissue, so that the mouth of the 'U' is towards the coelom. Usually the ends of the 'U' are closed, forming a box-like structure, but some species, e.g. *Eisenia foetida* have both

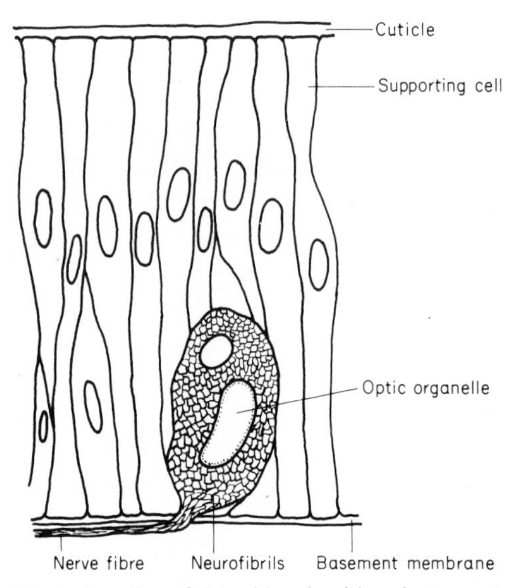

Fig. 6 Vertical section of the epidermis with a photoreceptor cell. (*After Stephenson, 1930*)

ends open, so that the bundles are disposed in radial columns. Between each bundle lies a double layer of connective tissue containing blood vessels. The inner surface of the longitudinal muscle layer is separated from the coelom, by a layer of coelomic epithelial cells, the peritoneum.

1.7 The coelom

The coelom is a large cavity that extends through the length of the body, and is filled with coelomic fluid (Fig. 7). It is surrounded on the outer side by the peritoneum of the body wall and on the inner

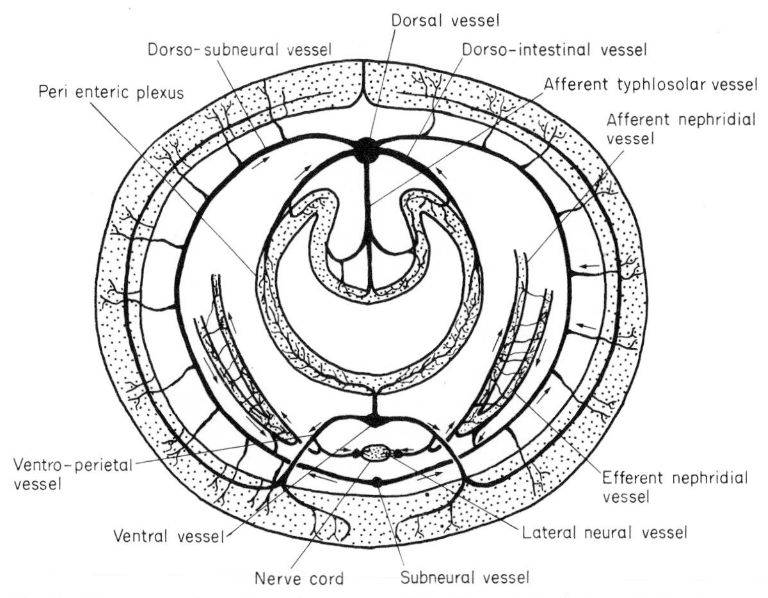

Fig. 7 Diagram to show the main segmental blood vessels in the intestinal region. (*After Grove and Newell, 1962*)

side by the peritoneum covering the alimentary canal. Transverse septa divide it into segmental portions. The peritoneum covering these septa is similar in structure to that covering the inner surface of the muscle layers; in a few species the peritoneum on the septa is so very much thickened that it almost fills the coelom in this region. The septa usually correspond to the external segmental grooves but often do not occur in the first few segments of the body, and sometimes they are missing in other parts, such as the oesophageal region of some species of *Pheretima.* Some species have septa that do not correspond with the intersegmental grooves in the anterior end but are displaced backwards, sometimes by as much as a segment. Two adjacent septa are sometimes fused at their junction with the body wall. The septa differ in thickness, depending on their position in the body, those in the anterior of the body being markedly thickened and more muscular. The degree of thickening, and the position and number of these septa are used as systemic features for many species of earthworm. Septa are constructed from muscle fibres, mostly derived from the longitudinal muscle layer, together with some circular muscles on the posterior face, with connecting tissue and blood vessels. The septa are perforated by pores which allow the coelomic fluid to pass freely between segments. Bahl (1919) showed that these pores are surrounded by sphincter muscles in species of *Pheretima,* and although Stephenson (1930) stated that sphincters are not present in other genera of the **Megascolecidae**, nor in *Lumbricus*, some later authors suggest that most earthworms have septa with at least one sphinctered opening.

There are also bands and sheets of mesenteric membranes between the body wall and the gut, forming pouches and dividing off some organs into separate chambers. Some peritoneal cells are modified in the region of the intestine as chloragogen cells, forming chloragogenous tissue. Chloragogen cells are characterized by the presence of yellow or greenish-yellow globules, the chloragosomes.

The coelomic fluid is a milky white liquid which is sometimes coloured yellow by eleocytes, cells containing oil droplets, as in *Dendrobaena subrubicunda.* The coelomic fluid of *E. foetida* smells of garlic, hence the name of this species. The consistency of the coelomic fluid differs between different species of earthworms, and also depends upon the humidity of the air in which the worms live; thus, it is thicker and more gelatinous in worms in dry situations than in those from wetter habitats.

The coelomic fluid contains many different kinds of particles in suspension. The inorganic inclusions are mainly crystals of calcium carbonate, but the corpuscular bodies in the coelomic fluid of lumbricid worms include phagocytic amoebocytes, feeding on waste materials; vacuolar lymphocytes (small disk-shaped bodies which do not occur in worms with large numbers of eleocytes) and mucocytes (lenticular bodies which give the coelomic fluid a mucilaginous component). Other inclusions in the coelomic fluid include breakdown products of the corpuscular bodies, protozoan and nematode parasites, and bacteria. 'Brown bodies' are aggregated dark-coloured masses or nodules usually found in the coelom at the posterior end of the body. They consist of disintegrated solid debris, such as setae and the remains of amoebocytes, and also cysts of nematodes and *Monocystis*.

Earthworms eject coelomic fluid through the dorsal pores, in response to mechanical or chemical irritation, or when subjected to extremes of heat or cold. Some species, such as *Megascolides australis*, eject fluid to a height of ten cm, and *Didymogaster sylvaticus* (known as the squirter earthworm) to a height of thirty cm. It has been suggested that this may discourage predators.

1.8 The alimentary canal

The alimentary canal or gut of earthworms is basically a tube extending from the mouth to the anus, although it is differentiated into a buccal cavity, pharnyx, oesophagus, crop, gizzard and intestine (Fig. 8). However, most aquatic worms do not have a gizzard (and sometimes lack a crop).

The short buccal cavity begins at the mouth, and occupies only the first one or two segments, with one or two diverticula or evaginations. The epithelium lining the buccal cavity is not ciliated, except occasionally in a small dorsal diverticulum, but very young worms do have patches of ciliated epithelium on the ventral surface. Overlying the epithelium is a thick cuticle, except in the ciliated portions.

The pharynx, which is not always obviously differentiated from the buccal cavity, extends backwards to about the sixth body segment. The dorsal surface of the pharynx is thick, muscular and glandular, and contains the pharyngeal glands, which appear as a whitish lobed mass. Some worms, for example, *Bimastos parvus*, have pharyngeal glands which extend back beyond the pharynx and are attached to the septa. Earthworms use the pharynx as a suction pump, so that muscular contractions of its walls draw particles through from the mouth, and some species evert the pharynx in this process.

All terrestrial oligochaetes have an oesophagus, which opens from the pharynx as a narrow tube, modified posteriorly as a crop and gizzard. Calciferous glands in evaginations or folds in the oesophageal wall open into the oesophagus. Lumbricidae have these folds most developed with their tips fused together, to form oesophageal pouches, which separate the glands from the oesophageal cavity.

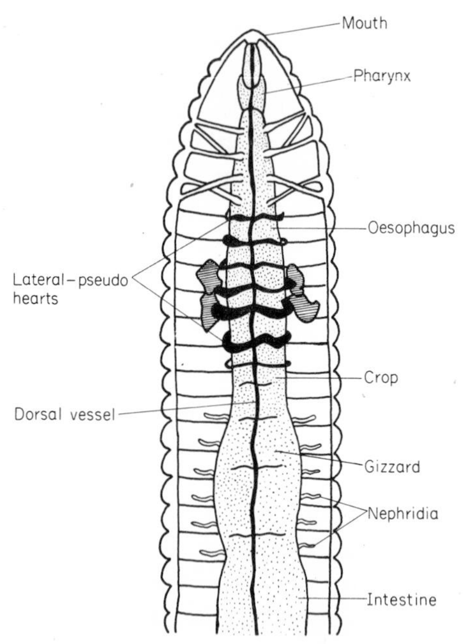

Fig. 8 Digestive and circulatory systems of *L. terrestris*. (*After Bachelier, 1963*)

At the posterior end of the oesophagus is the crop, a thin-walled storage chamber in front of the gizzard, which is muscular and lined with a thicker cuticle than the crop. Muscular contractions of the gizzard grind up the food with the aid of mineral particles taken in with the food. The crop and gizzard are much further forward in some of the other species of earthworms than in the Lumbricidae; they are immediately behind the pharynx in megascolecid worms which commonly have two to ten gizzards, each occupying one segment.

The rest of the alimentary canal is the intestine, which is a straight tube for most of its length, slightly constricted at each septum. Most of the digestion and absorption of food materials takes place in the intestine. The internal surface of the intestine has many small longitudinal folds and its surface area is increased by a large fold, the typhlosole, which projects from the dorsal wall (Fig. 7); this differs in relative size in different species, being largest in the lumbricid earthworms and completely absent in freshwater oligochaetes.

The epithelial lining of the intestine is composed mainly of glandular cells, and non-glandular ciliated cells. The ciliated cells of *Lumbricus* are also contractile, acting upon the gland cells which they surround, causing them to open or close. The intestine has two muscular layers, an inner circular and an outer longitudinal layer, the reverse order to that in the body wall.

1.9 The vascular system

There are three principal blood vessels, one dorsal and two ventral, that extend almost the entire length of the body, joined in each segment by blood vessels which ring the peripheral region of the coelom, and the body wall (Figs. 7, 8 and 9). The largest of the

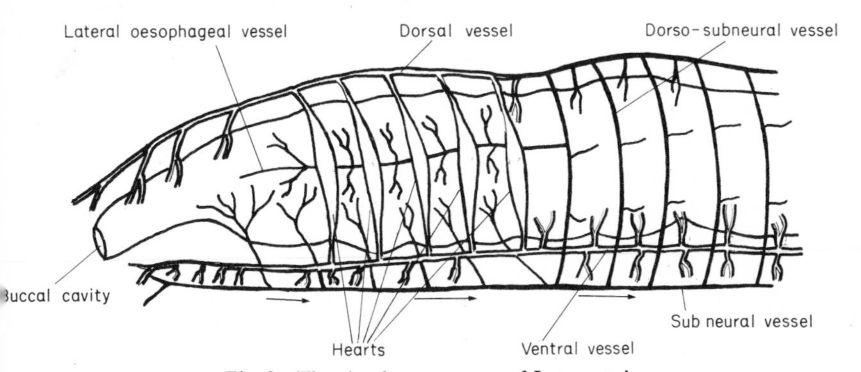

Fig. 9 The circulatory system of *L. terrestris.* (*Original*)

longitudinal vessels, the contractile dorsal vessel, is closely associated with the gut for most of its length, except in the most anterior portion, where it is separated from the gut by a mesentery. Some species of Megascolecidae and Glossoscolecidae have paired dorsal vessels for part or all of the length of the body. The ventral vessel, which is narrower than the dorsal vessel, lies immediately below the gut, and is suspended from it by a mesentery. The subneural vessel is even smaller than the ventral vessel. It lies beneath the ventral nerve cord, with which it is closely associated for its entire length. Some species do not have a subneural vessel, but most earthworms (except Megascolecidae) have two much narrower vessels which lie either side of the ventral nerve cord throughout its length (the latero-neurals). Megascolecid and some glossoscolecid earthworms, but not lumbricids have a supra-intestinal vessel (some generas, e.g. *Pheretima*, have two such vessels); this lies along the dorsal wall of the gut in the anterior segments and is part of the complex of blood vessels serving the alimentary canal. Other longitudinal blood vessels are the extra-oesophageal (or lateral-oesophageal) vessels which lie along either side of the gut, from the pharynx to the first of the dorso-subneural vessels.

The paired commissural vessels pass round the body in each segment from the dorsal vessel (and supra-intestinal vessel in those species which have it) to the ventral or subneural vessels. Some of the anterior commissures are enlarged, contractile, and with valves termed 'hearts' or 'pseudo-hearts'; *Lumbricus* has five pairs of such vessels, but some species have more and others less, e.g. megascolecid worms have between two and five pairs of 'hearts'. As well as these dorso-ventral anterior 'hearts', some species of earthworms have 'intestinal hearts' which connect the supra-intestinal vessels with the ventral vessel, and also other 'hearts' originating in branches from the dorsal and supra-intestinal vessels; *Pheretima* has 'hearts' of both these types. The 'hearts' direct the flow of blood through the body by means of valves similar to those in the dorsal vessel. The lateral-oesophageal-subneural commissures link the lateral-oesophageal and subneural vessels in the anterior segments of the body. Behind segment 12, a pair of dorso-subneural commissures, lying on the posterior face of each septum, link the dorsal vessel with the subneural vessel in each segment. A pair of ventro-parietal vessels on the anterior face of each segment branch off from the ventral vessel to the body wall and end in capillaries. Some Megascolecidae have supra-intestino-ventral commissures. The commissural or septal vessels are always associated with the faces of the septa, and sometimes the 'hearts' are attached by mesenteric-like folds of the peritoneum attached to the faces of the septa.

Between the alimentary epithelium and the outer muscular and peritoneal layers is a vascular network or plexus, the peri-enteric or alimentary plexus, which communicates with the dorsal and ventral vessels by two pairs of dorso-intestinal and three pairs of ventro-intestinal vessels per segment. The dorsal vessel supplies blood to the typhlosole via three small typhlosolar vessels per segment.

The body wall has capillaries which follow the muscle bundles and eventually connect with the subneural vessel via the lateral-oesophageal-subneural commissures in the front of the body, or via the dorso-subneural commissures in the rear. These capillaries also connect with the supra-intestinal vessel and the supra-intestino-ventral commissure behind the 12th segment of *Lumbricus*. The paired dorso-subneural commissures run between the dorsal and subneural vessels in every segment behind the 12th.

The efferent nephridial vessels join the dorso-subneural and the smaller and more numerous afferent nephridial vessels come from the ventro-parietals. The reproductive organs receive most of their blood supply from the ventro-parietal vessels.

1.10 The respiratory system

There are few specialized respiratory organs in terrestrial oligochaetes, although aquatic oligochaetes sometimes have gill-like organs. Most respiration is through the body surface which is kept moist by the mucus glands of the epidermis, the dorsal pores which exude coelomic fluid and the nephridial excretions through the nephridiopores.

There is a network of small blood vessels in the body wall of terrestrial earthworms, and oxygen dissolved in the surface moisture film permeates through the cuticle and epidermis to the thin walls of these vessels, where it is taken up by the haemoglobin in the blood and passed round the body. Species of *Lumbricus* also have looped capillaries extending from the ventro-parietal blood vessels into the epidermal layer (Fig. 10) (Stephenson, 1930). Combault (1909) stated that *Allolobophora caliginosa* f. *trapezoides* has special respiratory

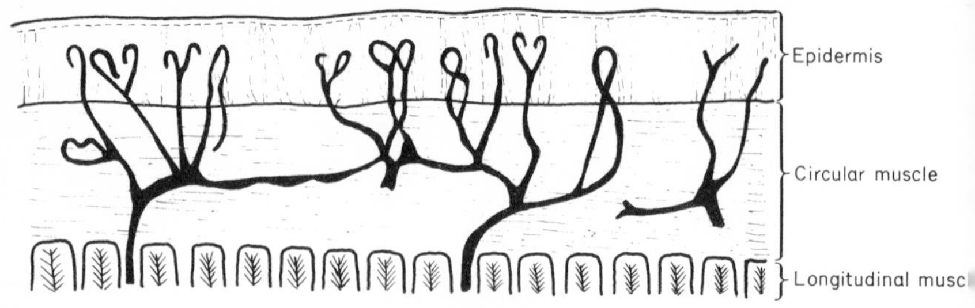

Fig. 10 Intra-epidermal capillaries of *L. terrestris*.
(*After Stephenson, 1930*)

regions of the body wall, where the cuticle is thinner, the epithelial cells shorter, there are fewer glandular cells in the epidermis, and there is a special subcutaneous blood supply.

1.11 The excretory system

The following description refers to *L. terrestris* and detailed differences in other species are discussed later. The nephridia, which are the main organs of nitrogenous excretion in oligochaetes, are paired in each segment except the first three and the last. The internal opening from the coelom into each nephridium is just in front of a septum, and is a funnel-shaped nephrostome (Fig. 11) which leads to a short preseptal canal that penetrates the septal wall into the segment behind, where the main part of the nephridium lies. The nephridium continues as a long postseptal canal with three loops, which can be distinguished as four sections: a very long 'narrow tube'; a shorter ciliated 'middle tube'; a 'wide tube'; and finally, the 'muscular tube' or reservoir, which opens to the exterior at the nephridiopore (Fig. 12).

The mouth of the nephrostomal funnel is flattened into two lips, an upper posterior lip and a lower anterior lip. The upper lip is much larger than the lower lip, and forms a semicircular expansion overhanging and projecting beyond the lower lip (Fig. 11). 'It may help the explanation to imagine that an ordinary conical funnel be compressed anterio-posteriorly, one lip much expanded and the lateral ends of the funnel depressed so as to increase the projection of the smaller lip' (Stephenson, 1930). The upper lip is composed of a radiating circle of marginal cells that are ciliated on their edges

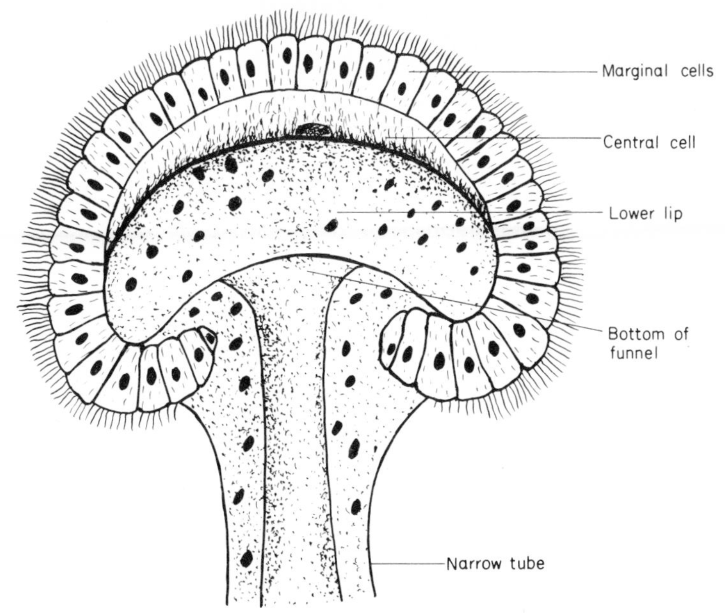

Fig. 11a Nephrostome – ventral view.
[*After Grove and Newell, 1962* (*based on Goodrich*)]

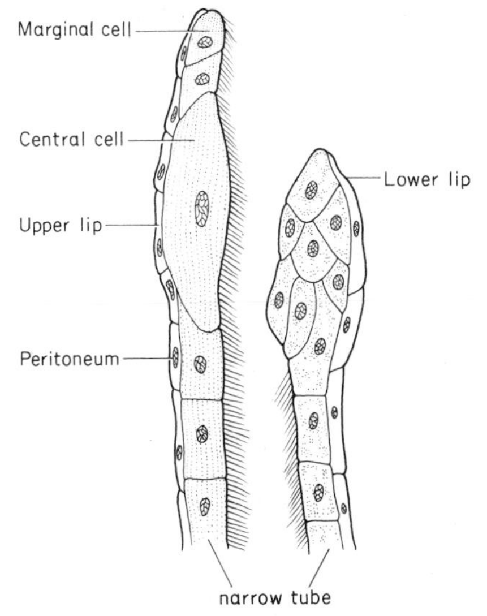

Fig. 11b Nephrostome – longitudinal section.
[*After Grove and Newell, 1962* (*based on Goodrich*)]

and inner faces, the cilia beating towards the canal. The lip is peripheral to a large crescent-shaped central cell with nucleate marginal cells and cilia on its lower face. The dorsal wall of the preseptal canal extends to the lower boundary of the central cell, forming a part of the upper lip which has no nuclei or cilia, neither has the lower lip which is much thinner than the upper lip. The short preseptal canal, which leads from the nephrostome to the wall of the associated septum, is intercellular, like most of the nephridial tubes,

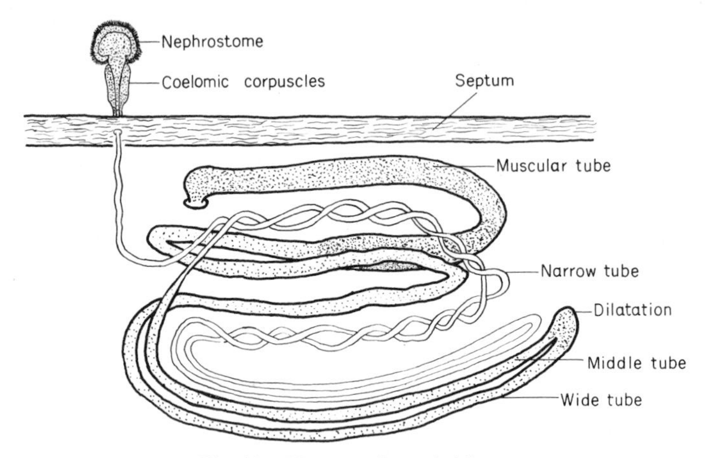

Fig. 12 Diagram of a nephridium.
[*After Grove and Newell, 1962* (*altered after Benham*)]

i.e. the lumen perforates a single column of cells which are called 'drain-pipe cells', and which have cilia in two longitudinal rows, one on each lateral wall.

Most of the post-septal part of the nephridial tube is secretory, except for the last portion (the muscular tube), and the lumen of the whole of the secretory section is intra cellular, also with 'drain-pipe cells'. The first part of the post-septal canal is the 'narrow tube', which has thin transparent walls containing granular protoplasm, nucleated at intervals. The only parts of this tube that are ciliated are at the beginning, about half-way along its length and at the end. After the narrow tube is the 'middle tube' which is shorter and wider, with thicker brown-coloured walls, and two rows of cilia along its whole length. This leads into the unciliated 'wide tube', which begins at a point where one of the loops bends sharply back on itself, via the ampulla, a dilated chamber with very large 'drain-pipe' cells. The inner surface of the ampulla is covered with a large number of rod-like bacteria (the function of these has been investigated by Knop (1926)) and the last part of the wide tube has a ringed appearance. The last part of the nephridial duct is the 'muscular tube' which is much wider than the preceding tubes, unciliated, and is the only part to have an intercellular lumen. This tube has a lining of flat epithelial cells surrounded by a muscular coat in its wall. It opens to the exterior by the nephridiopore through a sphincter muscle.

Nephridia of this pattern, i.e. opening externally on to the body surface (exonephric) occur in most lumbricids, but in *Allolobophora antipae*, all the nephridia from segment 35 backwards open into two longitudinal canals that discharge into the posterior part of the gut (enteronephric).

Large, paired nephridia such as described above, are known as meganephridia, but some earthworms also have micronephridia, which are much smaller and more numerous. For instance *Pheretima* has forty to fifty micronephridia attached to each septum except those septa in the anterior part of the body. Micronephridia are much simpler in form, and hang freely in the coelom from the ends of their funnels. Leading from a short narrow tube is the main part of the micronephridial tube, a long, spirally-twisted loop which has a ciliated lumen only in certain parts. The terminal ducts of these tubes pass into two septal excretory canals which curve around the posterior faces of the septa, and then pass to supra-intestinal excretory ducts that lie on either side of the dorsal surface of the gut. Some other megascolecid worms also have exonephric micronephridia.

Yet other kinds of nephridia end in a closed tube instead of a nephrostome; they are very small, exonephric, and attached to the body wall, with as many as 200–250 per segment. *Pheretima* has this type of nephridia as well as septal micronephridia and also tufted nephridia, which are bush-like organs consisting of bunches of nephridial loops on the walls of the alimentary canal in segments 4, 5 and 6, one pair per segment, the ducts from the loops joining up and emptying into the alimentary canal. Some species of *Megascolex* have tufted nephridia throughout the body; other species have them only in parts of the body. All earthworms have nephridia of one or more of the types described above, but the majority of species have open-funnelled exonephric meganephridia similar to those of *Lumbricus*.

The terms meganephridia and micronephridia are purely descriptive, but are disliked by some taxonomists who prefer the terms holonephridia and meronephridia which are not synonymous. Holonephridia are always single and relatively few in number in each segment. Meronephridia are multiple, although they are derived from a single rudiment in each segment. They are usually very numerous in each segment and much smaller than holonephridia.

1.12 The nervous system

The ventral nerve cord runs beneath the gut close to the ventral wall of the coelom, from the last segment of the body to the 4th segment from the front. Anteriorly, it passes into a suboesophageal ganglion,

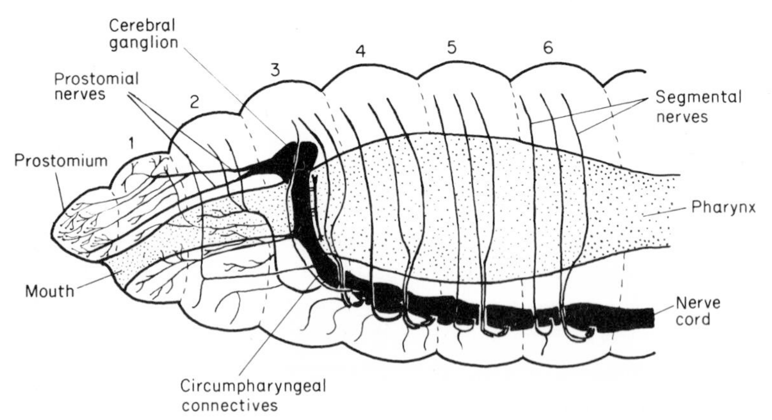

Fig. 13 Anterior part of the nervous system from the side. (*After Hess, 1925*)

then bifurcates into the circumpharyngeal (circumoesophageal) connectives, which pass up round either side of the oesophagus, and meet as the bilobed cerebral ganglion on the dorsal surface of the pharynx in segment 3 (Fig. 13). Structurally, the ventral nerve cord is really a backward extension of the two circumpharyngeal connectives, fused together. The ventral nerve cord is swollen into segmental ganglia in each segment, and from the 5th segment backwards, each of these ganglia has two pairs of segmental nerves branching from it in the posterior part of each segment, with a third pair just in front of each ganglion. These segmental nerves extend around the body wall, at first in the longitudinal muscle layer, and then in the circular muscle layer ending near the mid-dorsal line, thus forming almost a complete ring (some megascolecid species, e.g. *Pheretima*,

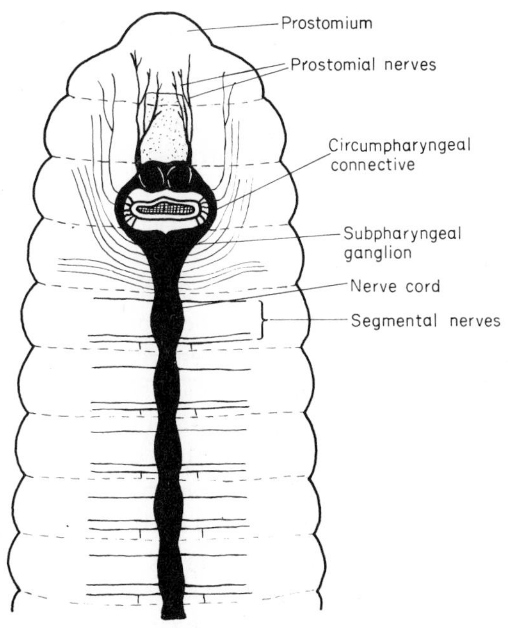

Fig. 14 Nervous system of *L. terrestris* from above. (*After Grove and Newell, 1962*)

have the ring completely closed). Each nerve ring has branches to the muscles, and to the epidermis. Each segmental nerve has a small subsidiary branch, close to the nerve cord, which passes down into the circular muscle layer, and then between this and the longitudinal muscle layer, ending in branches to the epidermis just before it reaches the mid-ventral line. Each septum is supplied by a pair of nerves (septal nerves) arising near the junction of the posterior segmental nerves and the ventral nerve cord (Fig. 14).

The distribution of the nerves in the anterior four segments differs from that in the other segments. Segment 3 has the typical distribution of three pairs of segmental nerves, but they originate from segment 4, which also possesses three pairs, so that six pairs of segmental nerves come from the ventral nerve cord in segment 4. Segment 2 is supplied with two pairs of nerves which originate from the junction of the circumpharyngeal connectives with the ventral nerve cord in segment 3, the larger posterior pair dividing, to give the normal three nerve rings in segment 2. The first segment is supplied by a pair of nerves originating in the lateral portions of the subpharyngeal connectives, which branch into two, shortly after they leave these nerves, and ramify in the first segment without forming nerve rings. A small nerve comes from the most ventral of these two branches to supply the ventral surface of the buccal cavity. The prostomium is innervated by two prostomial nerves, which originate from the front of the cerebral ganglion; these are the only nerves coming from this ganglion. A branch is given off from each prostomial nerve to supply the roof of the buccal cavity. The prostomial nerves then ramify through the prostomium, ending in the prostomial epithelium as nodule-like enlargements which then become part of the sub-epidermal nerve plexus. The last (caudal) segment of *Lumbricus* normally has six pairs of septal nerves instead of three (Hess, 1925), arranged as if they were in two successive normal segments. The most posterior of these six pairs of nerves are lateral terminations of the nerve cord itself.

The system of nerves supplying the gut is sometimes called 'the sympathetic system', and consists of a plexus of nerves which lie between the epithelium and the circular muscle layer of the alimentary canal throughout its length. Six small nerves from each circumpharyngeal connective unite in the external wall of the pharynx as a pharyngeal ganglion, from which nerve fibres pass forwards and backwards into the plexus. The muscles of the intestinal wall are supplied from a nerve plexus in the septum in each segment.

The wall of each blood vessel is covered with a network of nerve fibres and cells. Smallwood (1923, 1926) mapped the nerve supply to the major blood vessels, such as the dorsal and ventral vessels, but found no nerves to the hearts.

Before bifurcating, branches of the lateral nerves which link with the septal plexuses supply the longitudinal muscles of the body wall (Smallwood, 1926). The circular muscles are supplied by branches from nerve rings embedded in the body wall, which contribute to an inter-muscular nerve plexus. Between the basal membrane of the epidermis and the circular muscle layer is the sub-epidermal nerve plexus, similar to the sub-epithelial plexus in the wall of the alimentary canal. Nerve fibres from this plexus, which is linked with branches from the nerve ring, end between the epidermal cells, or enter the sensory cells.

The peritoneum has an extensive nerve net, derived from the nerves of the body wall.

The fine structure of the nervous system

The ventral nerve cord is covered externally by a peritoneum, and immediately beneath this is a layer of longitudinal muscle fibres within which lies a thin layer of tissue described as a neural lamella (Schneider, 1908). Beneath these layers, surrounding the nervous tissue, is the fibrous epineurium, which extends inwards into the nervous tissue and also outwards into the lateral nerves. This layer, which is ectodermal in origin, is made up of cells that resemble connective tissue. Three dorsally situated giant fibres, one median and two lateral, are separated from each other and from the rest of the nervous tissue by fibrous septa or lamellae. Stough (1926) stated that some *Lumbricus* species have two other smaller giant fibres in the ventral portion of the cord. There is also a double vertical septum dividing the cord into two lateral halves, which communicate in each segment by three large fenestrae through which pass fibrillae. The central portions of the nerve cord are further partitioned by extensions of the main horizontal and vertical septa. The nervous tissue in the cord is ramified by supporting tissue called the neuroglia which has both protoplasmic and fibrous cells, and a close network of fibres derived from processes of the cells. The rest of the central part of the nerve cord is occupied by the neuropile, a felt-like mat of fine anastomozing fibres derived from processes of the nerve cells. The nerve cells proper occupy the ventral and lateral portions of the nerve cord. They are termed bipolar or tripolar depending on the number of filamentous extensions, each cell always having more than one process (Stephenson, 1930), although they may appear monopolar if the two processes of a bipolar are so close together that they cannot be distinguished even after staining. The processes are distinguished as the axon, which is long, and one or more dendrites, which branch within the nervous system. Some of the cells of this region are large, pear-shaped and unipolar. These giant cells or ganglion cells link up with the giant fibres by long processes. The central portion of the nerve cord consists of a mass of fibrous material, forming felt-like masses with numerous perforations.

The cerebral ganglion, which contains no motor cells, is characterized by an external layer of small spindle-shaped ganglion cells, with a few large cells interspersed between them. The neuropile, which is very dense, is connected to the ganglion cells and contains the endings of the neural paths from the nerve cord. The histology of the main nerve rings has been described in detail (Dawson, 1920). Scattered along their length are nerve cells, which are most numerous on the posterior ring of each segment; bipolar nerve cells occur in the nerve plexus of the buccal cavity, unipolar 'ganglion cells' in the epithelium of the prostomium, and in the pharyngeal and intestinal plexuses.

The sense organs are of two types, photoreceptor organs and epithelial sense organs. The photoreceptor organs of lumbricids (Fig. 6) are 22–63 μ long, elongated and conical. Those occurring at the ends of the nerves to the skin are smaller than those in the epidermis (Hess, 1925) and are aggregated. One or sometimes two nerves pass into each cell, and large neural fibres also extend from the nerves into the cells. Small nerves continuous with the neural fibres extend throughout the protoplasmic contents of the cells as a neurofibrillar network. In each cell there is an ellipsoidal or elongated rod-shaped body, the optic organelle, with an outer surface consisting of interconnecting fibrillae, the retinella, and inside, a transparent hyaline substance.

The epithelial sense organs are groups of thirty-five to forty-five elongate cells, broader at their bases, with distal ends terminating in sensory hairs projecting through thin regions of the cuticle. The bases of the cells end as processes, one of which extends as a nerve fibre along the basal membrane and joins the nearest epidermal nerve. Proprioceptor cells situated in the muscle layers act as tension or stretch receptors.

1.13 The reproductive system

Oligochaetes are hermaphrodite, and have more complicated genital systems than unisexual animals. The reproductive organs, which are confined to comparatively few segments in the anterior portion of the body (Figs 15 and 16), include the male and female organs and associated organs, the spermathecae, the clitellum and other glandular structures.

The paired ovaries, which produce oocytes, are roughly pear-shaped in *Lumbricus* (or fan-shaped in *Pheretima*), and are attached by their wider ends to the ventral part of the posterior face of

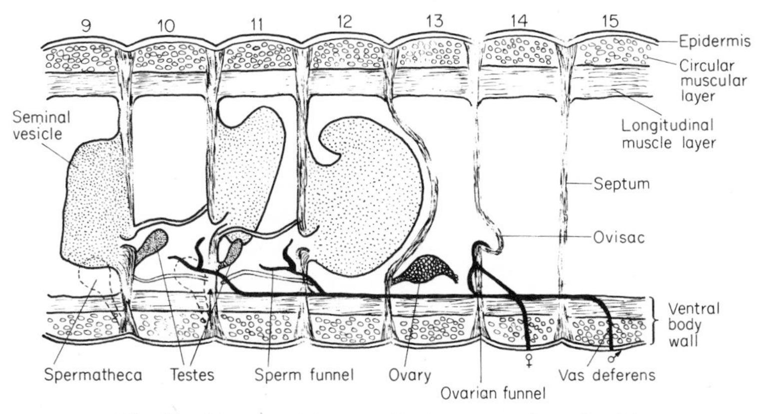

Fig. 15 Diagram of the reproductive system from the side. (*After Stephenson, 1930*)

septum 12/13, hanging freely in segment 13 in most terrestrial worms. The ovisacs are backward-facing evaginations of the anterior face of the septum immediately behind the ovaries and open into the dorsal wall of the ovarian funnels. They narrow posteriorly to form the oviducts, which in turn open on to the ventral surface of the body, their position depending on the family of earthworms.

The basic male organs are the testes. Most species of Lumbricidae, Megascolecidae and Glossoscolecidae have two pairs (holoandric) but some species of Lumbricidae and also the Ocnerodrilidae have only a single pair (meroandric). The testes are lobed organs attached to the posterior faces of septa 9/10 and 10/11 of *Lumbricus*, and projecting from the septal walls into two median testal sacs, one in segment 10 and one in segment 11. These testal sacs are separate compartments within segments 10 and 11, lying below the ventral vessel and enclosing the ventral nerve cord. Some species, e.g. *Pheretima* spp., have much more extensive sacs which contain the hearts, dorsal vessel, oesophagus and seminal vesicles. These sacs are filled with nutrient fluid in which lie the developing male cells. The testal sacs communicate with the seminal vesicles (vesiculae seminalis) which are storage sacs for the developing male cells.

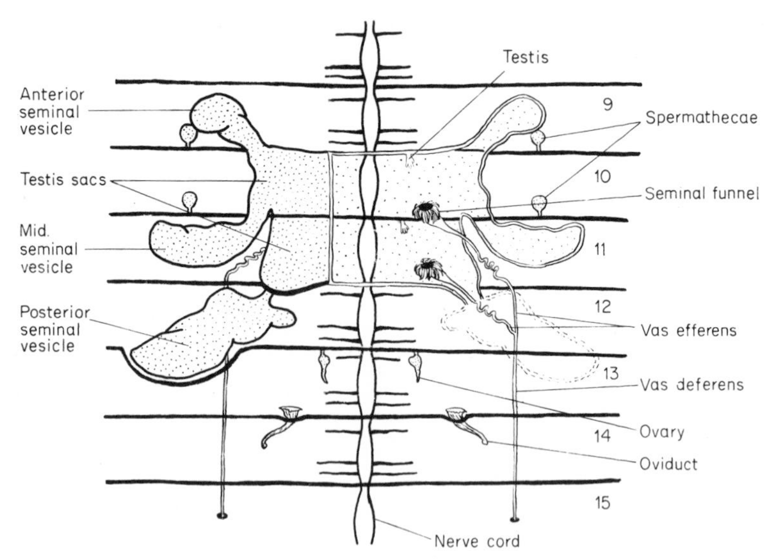

Fig. 16 Dissection of the reproductive system from above. (*After Grove and Newell, 1962*)

The seminal vesicles are the largest and most conspicuous organs of the reproductive system, and are immediately obvious in dissections as white masses on either side of the alimentary canal. They are sacs, for the most part divided by connective tissue into intercommunicating compartments. *Lumbricus* has three pairs of seminal vesicles in segments 9, 10 and 11, two pairs of which communicate anteriorly and posteriorly with the anterior testal sacs, and the third pair with the posterior testal sac in the segment immediately in front. Seminal vesicles are never in the same segment as the testes or the testal sacs. The seminal vesicles differ in size and are sometimes very large, extending backwards from the testal sacs through a number of segments, for example, they extend back as far as segment 20 in *Lumbricus castaneus*.

The sperm or funnel sacs (Fig. 15) are set into the posterior walls of the segments that contain them by their narrow portions. These sperm funnels are much folded, ciliated on their inner surfaces and open into coiled or straight male ducts, the vasa efferentia. The anterior and posterior vasa efferentia on each side pass backwards and join to

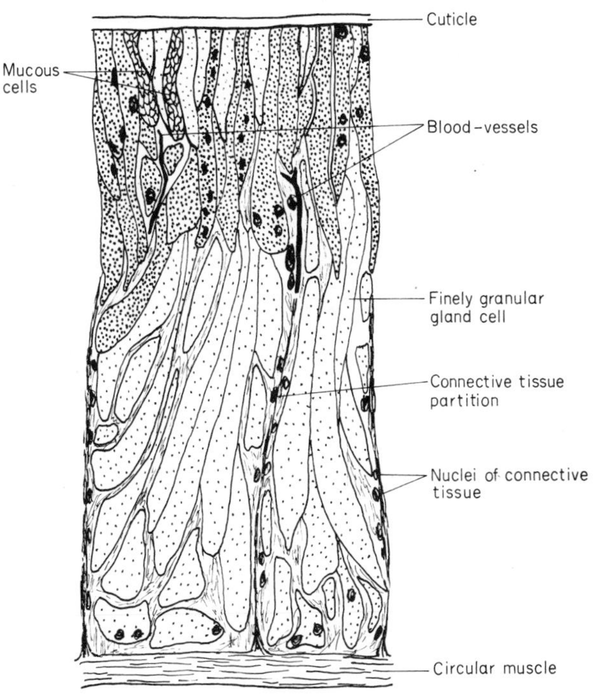

Fig. 17 Cross-section of clitellum.
(*After Stephenson, 1930*)

become a vas deferens, leading to the ventral exterior surface. Some species, for example *Pheretima posthuma*, have the ducts very closely associated, although not actually united, and each vas deferens opens to the exterior as a male pore.

Most oligochaetes have spermathecae, which are almost always paired organs: *L. terrestris* has two pairs, situated in segments 9 and 10, but other species may have more, up to a maximum of seven pairs, as in *Bimastos*, and some fewer than this or none. The spermathecae are attached to the body wall by short stalk-like ducts. Many species of megascolecids have one or more diverticula from the spermathecal ducts.

The clitellum consists of thickened glandular epidermis, particularly on its dorsal and lateral portions. A section through the clitellum (Fig. 17) shows gland cells in three layers; those nearest the surface are mucous cells, which are similar to the mucus-secreting goblet cells in the ordinary epidermis. Also reaching the surface, but extending deeper into the clitellar tissue, are long, slender and often convoluted cells, containing large granules. The bulk of the clitellar tissue in the deeper layer down to the basal membrane is made up of albumen-secreting gland cells terminating distally in long slender ducts which open on to the body surface. Epidermal columnar cells also occur. The posterior segments of the clitellum are less glandular than the anterior segments, with no cocoon-secreting and albumen-secreting cells, and fewer mucous cells.

Earthworms have glands known as 'prostates' associated with the male ducts. These have several different forms; those of the Megascolecidae are tubular or lobular while other families have muscular finger-like processes, or convoluted tubes. Prostates are rare in the family Lumbricidae.

2. Taxonomy

2.1 Systematic affinities and descent

Oligochaeta are grouped with the Polychaeta (bristle worms) and Hirudinea (leeches) in the Class Chaetopoda. Polychaetes are marine, whereas oligochaetes (except for a few species secondarily adapted) inhabit either soil or fresh water. Polychaetes may be considered as the older group because their larval development is more primitive than the oligochaetes, which have an embryo in a cocoon supplied by yolk or albumen. However, the greatest differences between the two orders are in the structure of their genital organs. Thus the genitalia of the Polychaeta, which have separate sexes, are simple, with extensive production of sexual cells from the coelomic epithelium, and sexual products that are expelled into the sea by rupturing the body wall. The oligochaetes, by contrast, are hermaphrodite, with their sexual organs confined to two or three segments and a very specialized and complex mechanism of fertilization and dispersal of eggs.

Possibly the Polychaeta are ancestral to the Oligochaeta, or they may both be derived from a common aquatic ancestor. Stephenson (1930) suggested that the Oligochaeta did evolve from the Polychaeta, but branched off before the modern families of the latter appeared. The Hirudinea possess characters that occur to some degree in many or all of the Oligochaeta, and are hence closely related to this order. Stephenson (1930) agreed with Michaelsen (1926) that the leeches were secondarily derived from the Oligochaeta, probably from the primitive aquatic family, Lumbriculidae.

Fossil records of oligochaetes are sparse, so there is little palaeontological information concerning the history and development of the order. The generic name *Protoscolex* (Ulrich) was used to describe four species of fossil segmented worms without setae or appendages that occurred during the upper Ordovician period in Kentucky, U.S.A.

Bather (1920) also described a species, ***Protoscolex latus***, from the upper Silurian period in Hertfordshire, England, which apparently bore papillae in one or two rows on each segment. He also placed ***Pronaidites***, another fossil worm which he considered to belong to the Oligochaeta, into this genus, suggesting that the papillae have some connection with setae. However, many authors are by no means convinced that these fossil worms are related to the oligochaetes, for instance Stephenson (1930) believed that the Oligochaeta were of comparatively recent origin, and that because of their feeding habits they arose no earlier than the Cretaceous era, when dicotyledenous plants appeared. Michaelsen (1910, 1922) thought that they evolved much earlier, at least as far back as the Jurassic period. As Benham (1922) found sporangia and vascular bundles of fern in the guts of some species of earthworms, he did not think that the appearance of dicotyledenous plants was necessary for the evolution of oligochaetes, although he believed that they may have increased considerably in numbers and species once dicotyledenous plants appeared. Arldt (1908, 1919) was convinced that the genus ***Acanthodrilus*** first occurred in the Triassic, ***Diplotrema*** in the lower Jurassic, ***Plutellus*** in the upper Jurassic and ***Megascolex*** in the Upper Cretaceous periods.

The early oligochaetes probably lived in mud rather than water (as presumably did their polychaete-type ancestor), becoming transiently terrestrial when the mud dried periodically. They then became gradually separated into two groups, one purely terrestrial, the other aquatic (in fresh water), so that some aquatic families such as Aeolosomatidae, Naididae and Tubificidae, have probably never passed through a terrestrial phase in their developmental history.

Stephenson (1930) thought that the common ancestor of the terrestrial Oligochaeta belonged to the aquatic Lumbriculidae, which is one of the most primitive of the oligochaete families. Of the modern families now described he considered the Moniligastridae to be the most primitive, and indeed, the structure of some members of this family is more similar to that of the aquatic worms than to that of any of the other terrestrial families. The Megascolecidae and

Eudrilidae have more advanced characteristics, but they still retain one primitive feature in the position of spermathecae and male or prostatic pores which come into contact during mating, thus allowing for the direct transfer of sperm. The Glossoscolecidae and Lumbricidae have fewest primitive features and may be considered to have evolved later than the other families, the Lumbricidae being most recent. Lumbricids never have a direct alignment of genital openings during copulation, so that a more elaborate mechanism is required to ensure effective transference of sperm.

2.2 Families, genera and species

Many authors have produced classifications of the Oligochaeta, but it was not until 1900 that Michaelsen produced the system that is the basis of the modern taxonomy of this group, and divided them into eleven families, containing about 152 genera and 1,200 species. Since then, Michaelsen (1921) reorganized his own classification into twenty-one families in two sub-orders, and Stephenson (1930) simplified this arrangement into fourteen families, which differed little from Michaelsen's original grouping. A division of families into the Microdrili, consisting of small, mainly aquatic worms (including the terrestrial Enchytraeidae) and the Megadrili (larger, mostly terrestrial worms) is now outdated, but may be useful in defining the scope of this book, which is concerned with terrestrial worms, and therefore does not include the Microdrili (although Enchytraeidae are terrestrial they are not dealt with here). The families that fall into this category are: *

(1) Moniligastridae.
(2) Megascolecidae (Ocnerodrilidae, Acanthodrilidae, Octochaetidae).
(3) Eudrilidae.
(4) Glossoscolecidae (Hormogastridae, Criodrilidae).
(5) Lumbricidae.

Of these families, the two most important are the megascolecid group and the Lumbricidae. The megascolecid group comprises more than half the known species, and includes worms that are very

* The families in brackets are those which Gates (1959) raised from Stephenson's original subfamilies, see page 43.

widely distributed outside the Palaearctic zone, with two genera, *Pheretima* and *Dichogaster*, that together probably contain more species than any other oligochaete genera. However, the most important family in terms of human welfare is undoubtedly the Lumbricidae, the most recently evolved family. This family of worms is of particular importance because it is the dominant endemic family in the Palaearctic zone including Europe, where until recent times many advances in agricultural practice have originated. Because of their ability to colonize new soils and become dominant to the near exclusion of local endemic species, the Lumbricidae have followed the spread of civilization, colonization and human development around the world. It has been said (Michaelsen, 1903) that only the genus *Pheretima* can successfully oppose an invasion by lumbricids. Thus, the earthworm populations in crop-growing areas are far more likely to consist mainly of species of the Lumbricidae than of members of any other family, and most work on the biology and ecology of earthworms has involved species of this family. Hence most of the discussions in this book concern members of the Lumbricidae.

It is not within the scope of this book to give more than a brief outline of earthworm taxonomy. The following list includes most of the known genera of earthworms, with an indication of their distribution, and the families to which they are currently assigned, together with a description of the family taxonomic features, taken from Stephensen (1930) unless stated otherwise.

2.2.1 Family MONILIGASTRIDAE

Setae. Single-pointed, sigmoid, four pairs per segment.

Male pores. One or two pairs in 10/11, 11/12 or 12/13.

Female pores. One pair, in 11/12 or on segments 13 or 14.

Spermathecal pores. One or two pairs, in 7/8 or 8/9 or 7/8 and 8/9.

Testes and funnels. One or two pairs, enclosed in one or two pairs of testal sacs suspended on the septa.

Vasa deferentia. Opening into prostate gland or independently on the surface.

Ovaries. One pair, with one pair of ovisacs extending backward from the ovarian segment.

Spermathecae. One or two pairs with long tubular ducts.

Gizzards. Multiple, either in front of the testis segment or segments or at the beginning of the intestine.
Nephridia. Meganephridial.

This family is usually considered the most primitive of the 'Megadrilid' families. The clitellum is only one cell thick and short male ducts reach the exterior surface in the segment behind that in which the funnels are situated. These characters are found only in this family and in the 'Microdrili'.

Distribution. Mainly South India. Also Ceylon, Burma, East Himalayas, Malay Archipelago, Philippines, Japan, China, Carolines Islands, Bahamas. (Subfam. Syngenodrilinae found only in tropical East Africa.)
Subfamily SYNGENODRILINAE
Syngenodrilus F. Smith and Green. Tropical East Africa.
Subfamily MONILIGASTRINAE
Desmogaster Rosa. Burma, Sumatra, Borneo.
Eupolygaster Michaelsen. Burma, Sumatra, Borneo.
Moniligaster Perrier. South India.
Drawida Michaelsen. South India and Ceylon, Burma, the East Himalayas; less abundantly in other parts of India and Pakistan, Borneo; peregrine species in the Carolines Islands, Sumatra, the Sunda Islands, Lombok, Java, the Philippines, Japan, China and the Bahamas.

2.2.2 Family MEGASCOLECIDAE

Setae. Single-pointed, either lumbricine or perichaetine in arrangement, sigmoid.
Male pores. Single pair, usually either on segment 17 or 18, seldom on segment 19.
Female pores. Paired or a single median pore, nearly always on segment 14.
Testes. Two pairs in segments 10 and 11, or one pair only in segment 10 or 11.
Prostates. One or two pairs, or rarely absent.
Ovaries. One pair in segment 13.
Gizzard. An oesophageal gizzard usually present.
Clitellum. Begins on or in front of segment 15.

Distribution. Mostly terrestrial; some species aquatic, a few littoral. Widely distributed throughout the whole of the southern hemisphere and southern part of the northern hemisphere. Not found in north and west Asia, northern Europe, and the north of North America. Only peregrine forms are found in mid and southern Europe and North Africa.

The classification of the megascolecid earthworms has always been more controversial than that of other oligochaete families, and in recent years three new systems of classification have been proposed, those of Omodeo (1958), Gates (1959) and Lee (1959), all of which would replace that of Stephenson (1930). The three systems are given in Table 1 (Sims, 1969). Omodeo recognized the taxonomic groups by the position and numbers of the calciferous glands, and on this basis raised one group to family status; Lee used the number and position of the male pores and position of the nephridiopores. Gates considered the structure of the prostatic glands and excretory system, and position of the calciferous glands important and raised all the main groups to family status. In an attempt to assess the relative merits of the three systems, Sims (1966) investigated the classification of a selection of megascolecid genera with computer techniques, by arranging them into groups with mutual characteristics using a dendrogram and vector diagram. Sims found that the pattern of the arrangement of the genera coincided to a large extent with the classification proposed by Gates, and he disagreed with those proposed by Lee and Omodeo. Therefore, for the purpose of this book, Gates' classification will be used. However, where it is more convenient, and may avoid confusion, the term 'megascolecid group' will be used for those genera in Stephenson's original Megascolecidae.

Gates (1959) split Stephenson's Megascolecidae into four families, Megascolecidae, Ocnerodrilidae, Acanthodrilidae and Octochaetidae, roughly corresponding with Stephenson's subfamilies, and Gates' descriptions and list of genera were as follows:

(*a*) Family MEGASCOLECIDAE (*sensu strictu*)

Prostates racemose, of *Pheretima* type, without a central canal, and presumably of mesodermal origin (Gates 1959).

TABLE 1

Three proposed reclassifications of Stephenson's (1930) Megascolecidae

Gates (1959)	Omodeo (1958)	Lee (1959)
Family MEGASCOLECIDAE Prostates lobular (racemose)	Family MEGASCOLECIDAE Prostates lobular (racemose)	Family MEGASCOLECIDAE as Stephenson (1930)
Family ACANTHODRILIDAE Prostates tubular Calciferous glands not in segments 9 or 9 and 10 Excretory system holonephridial	Family ACANTHODRILIDAE Prostates tubular Subfamily Ocnerodrilinae Calciferous glands in segment 9 or segments 9 and 10	Subfamily Megascolecinae One pair of prostatic pores combined with or in addition to one pair of male pores on segment 18
Family OCTOCHAETIDAE Prostates tubular Calciferous glands not in segments 9 or 9 and 10 Excretory system meronephridial	Subfamily proposed Calciferous glands absent Subfamily proposed Simple calciferous glands in segments 10–13	Subfamily Acanthodrilinae One pair of prostatic pores on segments 16, 17 or 19 or two pairs on segments 17 and 19 Tribe Neodrilacae Excretory system holonephridial Nephridiopores in two series alternating in position in successive segments
Family OCNERODRILIDAE Prostates tubular Calciferous glands in segments 9 or 9 and 10	Subfamily proposed Simple calciferous glands in segments 14–17 Subfamily Benhaminae Calciferous glands stalked and laminate	Tribe Acanthodrilacae Excretory system holonephridial with nephridiopores in a single series on each side of the body; or excretory system meronephridial

(*adapted from Sims 1969*)

Genera

Pheretima Kinberg. The most widely distributed genus of the megascolecid group. Indo-Malaya, eastern Asia, Australia and many peregrine species distributed throughout the world.
Perionyx Perrier. India, Ceylon, Australia, New Zealand.
Woodwardiella Stephenson. India, Ceylon, Australia, Java.
Lampito Kinberg. India.
Didymogaster Fletcher. Australia, New Zealand.
Notoscolex Fletcher. India, Ceylon, Australia, New Zealand.
Pliconogaster Beddard. Australia (South and West), New Zealand, South India.
Comarodrilus Stephenson. India.
Megascolex Templeton. India, Malaya, Australia, New Zealand, Pacific Islands.
Digaster E. Perrier. Australia (Queensland, N.S.W.).
Perissogaster Fletcher. Australia (Queensland, N.S.W.).
Nellogaster Gates. India, South-east Asia.
Nelloscolex Gates. India, South-east Asia.
Exxus Gates. Australasia.
Neochaeta Lee. New Zealand.

(*b*) Family OCNERODRILIDAE

Pre-intestinal region short, intestinal origin in or anterior to segment 15, with latero-oesophageal hearts confined to segments 10–11. Setal arrangement lumbricine. Calciferous glands or epithelial-lined diverticular spaces in thickened oesophageal wall, in segments 9–10. Excretory system holonephric (Gates 1959).

Genera

Maheina Michaelsen. Seychelles.
Curgiona (*Curgia*) Michaelsen. South India. (Aquatic.)
Malabaria Stephenson. South India (Aquatic or amphibious.)
Quechua Michaelsen. Peru.
Paulistus Michaelsen. Brazil.
Eukerria (*Kerria*) Beddard. Subtropical South America (Brazil, Paraguay) Lower California, West Indies (St Thomas Island). Peregrine species in South Africa and New Caledonia.

Kerriona Michaelsen. Eastern Brazil.

Haplodrilus Eisen. Paraguay, Brazil.

Ocnerodrilus Eisen. America from California to Guyana and Colombia, West Indies, tropical and southern Africa. Some species peregrine. (Some species aquatic.)

Pygmaeodrilus Michaelsen. Tropical East and Central Africa, South-west Africa. (Aquatic or in marshy ground.)

Nematogenia Eisen. Southern Nigeria, Liberia, Costa Rica. One species peregrine in Bahamas, Panama, Ceylon.

Gordiodrilus Beddard. Tropical and subtropical Africa, Madagascar, South India.

Deccania Gates. India.

(*c*) Family ACANTHODRILIDAE

Pre-intestinal region longer than in Ocnerodrilidae, with intestinal origin in or behind segment 15, and with hearts not confined to segments 10–11 or their homoeotic equivalents. Excretory system holonephric.

Genera

Acanthodrilus E. Perrier em Michaelsen. New Caledonia, Australia, New Zealand, Mexico, Cuba, Guatemala, Chile, southern Argentine, South and South-west Africa, Madagascar.

Microscolex Rosa em Michaelsen. Southern South America and islands, South Africa and sub-Antarctic islands.

Rhododrilus Beddard em Michaelsen. New Zealand and associated islands, Australia (Queensland).

Dinodriloides Benham. New Zealand.

Leptodrilus Benham. Auckland and Campbell Islands.

Periodrilus Michaelsen. New Zealand.

Maoridrilus Michaelsen. New Zealand.

Neodrilus Beddard. New Zealand.

Plagiochaeta Benham. New Zealand.

Chilota Michaelsen. Southern South America and islands, South Africa, Cape Verde Islands.

Yagansia Michaelsen. Southern South America (Andes) and southern islands.

Udeina Michaelsen. Local areas in South Africa.
Parachilota Gates. India.
Diplocardia Beddard. United States, Mexico, India.
Zapotecia Eisen. Mexico, Haiti.
Plutellus E. Perrier. India and Ceylon, Australia, Tasmania, New Caledonia, Auckland Islands, North America.
Diplotrema Spencer. Australia (Queensland), New Caledonia.
Pontodrilus E. Perrier. Found chiefly on the sea shore, and is very widely distributed. Islands of the Southern Hemisphere, islands and coasts of North America, New Zealand and Ceylon. The coasts of Sardinia and southern France.
Eodrilus Michaelsen. Distribution as for *Acanthodrilus*.
Diprochaeta Beddard. (Australia, Queensland, Victoria, Tasmania) New Zealand and islands, India. (One aquatic species.)

(*d*) Family OCTOCHAETIDAE

Description as for Acanthodrilidae except excretory system meronephric (Gates, 1959).

Genera

Howascolex Michaelsen. South India, South Madagascar.
Octochaetus Beddard. India, New Zealand.
Deinodrilus Beddard. New Zealand.
Hoplochaetina Michaelsen. New Zealand.
Ramiella Stephenson. India.
Eudichogaster Michaelsen. India.
Eutyphoeus Michaelsen. India.
Hoplochaetella Michaelsen. West and South India.
Trigaster Benham. Mexico, West Indies (St Thomas Island).
Dichogaster Beddard. America (North, Central, South) from California to Ecuador, and West Indies, tropical Africa, India (possibly one or two endemic species). Peregrine species in Brazil, Paraguay, southern Asia, Malay Archipelago, Polynesia, North-west Australia, Madagascar.
Megascolides McCoy. India, Australia. One species in North America.
Spenceriella Michaelsen. Australia (Victoria), New Zealand (Little Barrier Island), South India.

Eutrigaster Cognetti. Costa Rica.
Monogaster Michaelsen. Southern Cameroon.
Leucodrilus Lee. New Zealand.
Lennogaster Gates. India.
Benhamia Michaelsen. North and Western Africa, Congo.
Millsonia Beddard. Guinea, Nigeria.
Wegeneriella Michaelsen. Cameroon, tropical South America.
Aeogaster Gates. India.
Scolioscolides Gates. India (a single species).
Barogaster Gates. India.
Rillogaster Gates. India.
Priodochaeta Gates. India.
Priodoscolex Gates. India.
Travoscalides Gates. India.
Celeriella Gates. India.
Tonoscolex Gates. India.

2.2.3 Family EUDRILIDAE

Setae. Eight per segment.

Male pores. Single or paired on segments 17–18.

Spermathecal pores. Single or paired, on or further back than segment 10.

Prostates. Organs termed 'euprostates' which are neither tubular nor lobular, but terminations of the vasa deferentia, often strongly muscular, appear as cylindrical or finger-shaped projections into the body cavity.

Nephridia. Meganephridial.

Spermathecae. Very closely associated with the other female organs (ovaries, oviducts, funnels and egg sacs) often connected directly or by coelomic tubes and sacs).

Distribution. Except for one or two species, confined to tropical and subtropical Africa.

Subfamily PARENDRILINAE

Genera

Platydrilus Michaelsen. North-east Africa, Kenya, Tanzania, Congo, Rhodesia.

Eudriloides Michaelsen. Tropical and southerly subtropical East Africa.
Megachaetina Michaelsen. Tanzania.
Borgertia Michaelsen. Tanzania, Zanzibar Island.
Notykus Michaelsen. Tanzania.
Metadrilus Michaelsen. Tanzania.
Libyodrilus Beddard. Cameroon.
Parendrilus Beddard. Equatorial Central and East Africa.
Nemantodrilus Michaelsen. Mozambique, South-west Africa.
Chuniodrilus Michaelsen. Liberia.

Subfamily EUDRILINAE

Genera

Eminoscolex Michaelsen. North-east Africa, Cameroon, equatorial Central Africa.
Gardullaria Michaelsen. North-east Africa.
Nenmanniella Michaelsen. North-east Africa, equatorial Central and East Africa.
Bettonia Beddard. Tropical Africa.
Teleudrilus Rosa. North-East Africa, Tanzania.
Polytoreutus Michaelsen. Equatorial East and Central Africa.
Eupolytoreutus Michaelsen. Central Africa, North-east Cameroon.
Schubotziella Michaelsen. Congo.
Eutoreutus Michaelsen. North Nigeria, Congo.
Büttneriodrilus Michaelsen. Equatorial West Africa, Congo.
Rosadrilus Cognetti. Cameroon.
Beddardiella Michaelsen. Nigeria, Cameroon.
Malodrilus Michaelsen. Ethiopia, Cameroon.
Eudrilus E. Perrier. West Africa.
Kaffania Michaelsen. Ethiopia.
Euscolex Michaelsen. Cameroon.
Metascolex Michaelsen. Cameroon.
Parascolex Michaelsen. Cameroon, Togoland.
Hyperiodrilus Beddard. Togoland, South Nigeria, Congo.
Iridodrilus Beddard. Cameroon.
Legonea Clausen. Ghana.

2.2.4 Family GLOSSOSCOLECIDAE

Setae. Mostly single-pointed, rarely double-pointed, usually ornamented, sigmoid. With few exceptions, 8 per segment. Penal setae absent, copulatory setae often present.

Male pores. In clitellar region, or in front, rarely behind.

Dorsal pores. Absent.

Vasa deferentia. Ectal and usually simple, however often with copulatory pouch, rarely with prostatic glands.

Clitellum. Usually beginning behind segment 14.

Gizzard. Usually single, rarely multiple, in front of the testis segments, often with one rudimentary gizzard at the hinder end of the oesophagus, behind the ovarian segment.

Nephridia. Meganephridial.

Distribution. Mostly terrestrial forms, but a number are littoral and several are found in fresh water. Terrestrial and littoral forms occur in America (Mexico to Argentina), West Indies, southern Europe, Malagasy and the southern portion of Africa. Freshwater forms occur in North and South America (except the colder parts of the temperate zones), Europe, South-west Asia, the Malay Archipelago and the warmer parts of Africa.

Stephenson's subfamilies have in some instances been raised to family status by some recent authors, and this is noted in the following list of genera.

Subfamily GLOSSOSCOLECINAE

Genera

Thamnodrilus Beddard. Central and South America (Panama, Colombia, Peru, Ecuador, Guyana).

Aptodrilus Cognetti. South America (Columbia, Ecuador, North Brazil).

Rhinodrilus E. Perrier. South America (Columbia, Venezuela, Brazil, perhaps also Argentina and Paraguay).

Andiorrhinus Cognetti. South America (Venezuela, North Brazil, Bolivia, Paraguay).

Onychochaeta Beddard. Bermudas, West Indies, South America (Venezuela, Guyana).

Pontoscolex Schmidt. Central America, South America (Colombia, Guyana, Ecuador, Brazil and near the coast or on islands in the tropical belt).

Opisthodrilus Rosa. South America (Brazil, Paraguay, Argentina).

Anteoides Cognetti. South America (Bolivia, Argentina, North Paraguay).

Periscolex Cognetti. Central and South America (Panama, Columbia, Ecuador).

Holoscolex Cognetti. Ecuador.

Enantiodrilus Cognetti. Argentina.

Glossoscolex Leuck. Tropical South America, especially the eastern part (Minas Gẽraes to Rio Grande do Sul, one species in Ecuador).

Andioscolex Michaelsen. Southern Central America (Darien) and tropical South America, especially the Andes region (from Colombia to Ecuador, perhaps also to Bolivia), one species in Brazil (Province of Rio de Janeiro).

Fimoscolex Michaelsen. Brazil.

Subfamily SPARGANOPHILINAE

Genus

Sparganophilus Benham. North and Central America (U.S.A., Mexico, Guatemala).

Subfamily MICROCHAETINAE

Genera

Microchaetus Rapp. South America.

Tritogenia Kinberg. South Africa.

Glyphidrilus Horst. Tanzania, India, Burma, the Malay Peninsula, the Malay Archipelago, Indonesia.

Kynotus Michaelsen. Madagascar.

Callidrilus Michaelsen. Tanzania, Mozambique.

Drilocrius Michaelsen. Tropical South and Central America.

Alma Grube. Central and North-east Africa (lower Egypt, Sudan, West Uganda, Tanzania, Rhodesia (Middle Zambesi) Congo, Cameroon, Nigeria, Togoland, Gambia.

Subfamily HORMOGASTRINAE; family HORMOGASTRIDAE (Gates)

Hormogaster Rosa. Southern Europe and North Africa, Italy, Corsica, Sardinia, Sicily, Tunis, Algeria.

Subfamily CRIODRILINAE (family CRIODRILIDAE (Gates))

Criodrilus Hoffmeister. Germany, Austria, Hungary, Italy, South Russia, Syria, Israel, India, Japan.

2.2.5 LUMBRICIDAE

Setae. Single-pointed, lumbricine arrangement, often ornamented, sigmoid. Copulatory setae on certain anterior segments ridged and grooved on their distal portions, and usually on raised papillae.

Male pores. Usually on segment 15, or rarely displaced, one to four segments forward.

Female pores. Usually on segment 14.

Testes. Two pairs, and two pairs of funnels in segments 10 and 11 with no prostates, but rarely, prostate-like glandular cushions are present.

Spermathecae. If present, simple, without diverticula.

Ovaries. In segment 13.

Gizzard. Single, well-developed, at the beginning of the intestine, oesophagus possessing calciferous glands.

Clitellum. Saddle-shaped, beginning behind the male pores.

Distribution. Mostly terrestrial, a few in fresh water. Temperate and colder regions of the Northern Hemisphere; Japan, Siberia, Central Asia, Europe, North India and Pakistan, Israel, Jordan, North America. Many peregrine species all over the world.

Genera

Eiseniella Michaelsen. Europe, Syria and nearby areas, Azores, Canary Islands, North America (except Mexico), Chile, South Africa, Australia (N.S.W.), New Zealand. (Marshy land or aquatic.)

Eisenia Malm. Siberia, southern Russia, Israel, Europe, North America.

Dendrobaena Eisen. South and West Siberia, northern India, Syria, Israel, Caucasus, Europe, Iceland, Greenland, North America, Azores, southern South America, Egypt.

Allolobophora Eisen. Japan, northern India, Pakistan, Persia, Syria, Israel, North Africa, Europe, Canary Islands, North America.

Eophila (*Helodrilus*) Rosa. Caucasus, Europe, Israel, Syria, northern India, Pakistan.

Bimastos Moore. Asia Minor, northern India, Pakistan, Europe, Azores, North America, South America (except tropical areas), South Africa, Hawaii. (Sometimes semi-aquatic.)

Octolasium Oerley. Caucasus area, Europe, North Africa, Syria, Azores, North America, Australia (N.S.W.) and transported to many other places. (Often semi-aquatic.)

Lumbricus Linnaeus. Siberia, Europe, Iceland, North America. Transported throughout the world.

The characters used to define genera of the Lumbricidae are mostly internal structures, such as the number and distribution of the organs of the reproductive system, such as testes sacs, spermathecae and seminal vesicles, and also the position of the gizzard. Externally, the positions of the male pores, spermathecal and first dorsal pores and the spacing of the setae are generic and specific characters. Other features that are used when describing a genus fully, include cephalization (mode of attachment of prostomium to the peristomium), position of the female pores, the presence or absence of pigment and the segmental position of the clitellum.

There are about 220 species (Cernosvitov and Evans, 1947) of which nineteen are common in Europe, and have spread throughout the world, mostly through the agency of man. External characters such as cephalization, the position and extent of the clitellum and tubercular pubertatis, the position of the first dorsal pore, the prominence of papillae, the position and spacing of genital setae, the spacing of setae, colour, length of body and number of segments of the adults are all used in describing species. Some species have one or more variants, forms, or subspecies which differ in only one or two very minor characters from the typical form, and sometimes the variant replaces the typical form in a particular area. The number of species in the family Lumbricidae is small compared with that of other families, and diagnostic differences between genera and species are not great. This family is still being studied taxonomically and further divisions, particularly at species level, may be recognized in the future.

2.3 Geographical distribution

Earthworms occur all over the world, but only rarely in deserts, areas under constant snow and ice, mountain ranges and areas almost entirely lacking in soil and vegetation. Such features are natural barriers against the spread or migration of earthworm species, and so are the seas, because most species of earthworms cannot tolerate salt water even for short periods. Nevertheless, some species of earthworm are widely distributed, and Michaelsen has used the term 'peregrine' to describe such species, whereas the other species that do not seem able to spread successfully to other areas to any great extent have been termed 'endemic' species.

The most common and versatile of the peregrine lumbricid species have spread to many areas, especially those places that have been colonized from Europe, so that in many sites close to human habitation the European species of earthworms have become dominant and almost eliminated the local species. For instance, in Chile, New Zealand, areas of the United States, South-west Africa, North-west India and Australia, one of the two commonest species is now *Allolobophora caliginosa*.

Megascolecid earthworms of the genus *Pheretima*, which are indigenous to South-east Asia, have also migrated to many tropical, subtropical and sometimes even temperate regions. Several peregrine species occur in South and Central America and the West Indies, and three species are widely distributed in India. Other peregrine species include *Dichogaster*, *Pontoscolex corethrusus* (Glossoscolecidae), and *Eudrilus eugeniae* (Eudrilidae) in many parts of the tropics, and *Microscolex phosphoreus* (Megascolecid group) which is common in many parts of the world.

Certain areas can be characterized by their endemic species. The area that includes Europe, and Asia north and west of the Himalayas

(Palaearctic zone) is characterized by the Lumbricidae. The northern boundary of endemic species of lumbricids is the limit of the ice-age glaciation, and further north than this (in an area which includes all but the most southern parts of England and Ireland) usually only peregrine species are found. Probably glaciation exterminated the endemic species in these areas, and they were repopulated by peregrine species when the glaciers retreated. The only other endemic earthworms in the Palaearctic zone are a group of earthworms which includes the genus *Hormogaster* (Hormogastridae), one species attaining a length of 75 cm (*Hormogaster redii* f. *gigantes*). Members of this genus are found in Sardinia, Corsica, Italy, Sicily, southern France and North Africa. The northern part of North America, particularly Canada and Alaska, has no endemic earthworms, probably because of glaciation. The United States has a few endemic species of lumbricids in its eastern region, but most species of this family in North America have been introduced to the continent by man. Some megascolecid earthworms, of the genera *Plutellus* and *Megascolides*, occur west of the Rocky Mountains, while the acanthodrilid genus *Diplocardia* occurs throughout the United States.

Endemic species of the subfamily Glossoscolecinae dominate the New World south of Mexico, Trinidad and Tobago, and as far as the River Plate in the south-east, although not so far south in the west. In Mexico and most of the West Indies and Central America, a number of peregrine species of Glossoscolecinae occur. However, most of the earthworms in this area belong to the genera *Zapotecia*, *Trigaster* and *Dichogaster*, *Acanthodrilus* and *Plutellus*, all of which are megascolecid group genera. The temperate southern portion of South America has a separate earthworm fauna, predominantly *Microscolex*.

Three regions of Africa have characteristic species of earthworms; these are the tropical region from the southern border of the Sahara in the north to the tropic of Capricorn in the south; the southern part of the continent including the Cape of Good Hope; and Malagasy and neighbouring islands.

Two groups of earthworms predominate in the tropical area of Africa, the family *Eudrilidae* and a number of species of *Dichogaster* (Acanthodrilidae). The Eudrilidae are endemic but the genus

Dichogaster also occurs in Central America and the West Indies. Some species of Ocnerodrilidae are also found in this region, of which one genus, *Pygmaeodrilus*, is confined to Africa. The southern part of Africa has two groups of endemic worms, the Microchaetinae, a subfamily of the Glossoscolecidae, and earthworms of the family Acanthodrilidae.

East and South-east Asia (including India), Australasia and the Pacific Oceanic Islands are dominated by earthworms of the megascolecid group, with at least ten genera in India and Pakistan, and at least thirty species of *Megascolex*, and about a dozen species belonging to three other megascolecid genera in Ceylon.

The area that includes South-east Asia, China and Japan is characterized by the megascolecid genus *Pheretima*; this species is also found extensively in India. There are several other genera of megascolecids, particularly in the Indonesian islands, and some species of moniligastrids occur in Burma. Japan has a single endemic lumbricid species. The endemic earthworms of Australasia are megascolecids, including some endemic species of *Pheretima*, although most species of this genus in this area are undoubtedly peregrine.

Earthworms of the family Acanthodrilidae which are not found in South-east Asia, occur in Australia, New Caledonia and particularly New Zealand (the genus *Deinodrilus* is confined to New Zealand), and also India and Ceylon. Many megascolecid and acanthodrilid genera are found only in Australasia.

A few earthworms (species unknown) have been reported from the South Shetland Islands, in Antarctica.

3. Biology

3.1 Life cycles

It should be emphasized that our present knowledge of the life cycles of even quite common earthworms is very inadequate, and requires much more study.

The ova of earthworms are contained in cocoons (oothecae), which most lumbricids leave near the soil surface. If the soil is very moist, they will deposit them almost on the surface, but if it is very dry they are placed much deeper. Many species of earthworms produce cocoons throughout the year, when the temperature, soil moisture, food supplies and other environmental factors are suitable, but it has been demonstrated clearly in cultures (Evans and Guild, 1948), and in the field (Gerard, 1967) that seasonal fluctuations of the soil climate also cause the number of cocoons produced by different species of earthworm to vary.

Evans and Guild showed that in cultures of common species of lumbricids, kept for a year at what they considered to be optimum soil moisture, the number of cocoons produced by each species closely paralleled the seasonal changes in soil temperature (Fig. 18). Fewest cocoons were produced in the winter months and there was a temperature threshold of about 3°C, below which no cocoons were produced. Many cocoons were produced from the end of February to July when temperatures were rising, and the greatest numbers occurred between May and July. Thereafter, the numbers of cocoons produced decreased quite rapidly with the falling temperatures, particularly for some species. Gerard (1967) confirmed these conclusions by sampling in the field the cocoons of *Allolobophora caliginosa* and *Allolobophora chlorotica* which were also studied by the other authors. In his study, most cocoons occurred in samples taken in May and June; probably more cocoons were produced in July and later, but at this time the cocoons would be hatching faster than they were being produced, so the numbers found in soil samples would decrease. Very few cocoons occurred in samples taken from August to November; although some were found in the

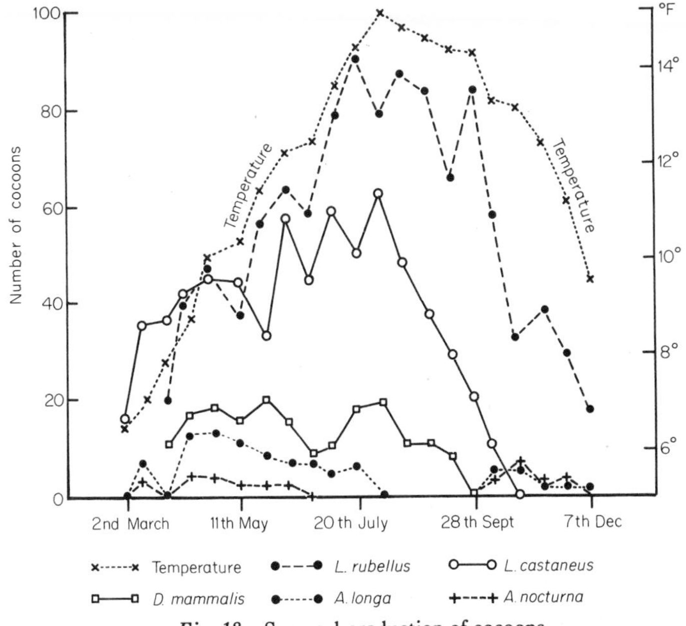

Fig. 18 Seasonal production of cocoons.
(*After Evans and Guild, 1948*)

December, January and February samples, these were probably an accumulation because cocoons produced during the colder months would not hatch until the soil warmed up in the spring.

Lumbricids such as ***Allolobophora longa*** and ***Allolobophora nocturna*** have an obligatory diapause during the summer months, and so produce cocoons only from mid-March to June or early July, and October to November. Other lumbricids belonging to species of the genera ***Eisenia*** and ***Octolasium*** and other species of ***Allolobophora*** may have a facultative diapause during dry periods, and this also interrupts the production of cocoons.

The number of cocoons produced in a season varies very greatly with species and climate. Table 2 shows the number of cocoons produced by individuals of eleven different species in culture. Satchell (1967) pointed out that there was a striking correlation between the number of cocoons produced by any species and how much this species is exposed to adverse environmental factors such as desiccation, extremes of temperature and predation. In other words, species of earthworms that may be exposed to environmental hazards normally produce more cocoons to enable them to survive adverse conditions. Thus, those species that live or can move into the deeper soil layers and are protected from adverse conditions, e.g.

TABLE 2

Number of cocoons produced per annum by one individual

	Year	No. of cocoons		Year	No. of cocoons
L. rubellus	1946	79	*O. cyaneum*	1946	13
L. castaneus	1945	65	*E. foetida*	1946	11
D. subrubicunda	1946	42	*A. rosea*	1946	8
A. caliginosa	1946	27	*A. longa*	1945	8
A. chlorotica	1946	25	*A. nocturna*	1945	3
D. mammalis	1945	17			

(*From Evans and Guild, 1948*)

Lumbricus terrestris, *Allolobophora longa* and *Octolasium cyaneum* normally produce fewest cocoons, whereas those species that live near the surface, such as *Lumbricus castaneus*, *Lumbricus rubellus* and *Dendrobaena subrubicunda*, and hence are exposed more directly to these factors, produce very many more. As individuals of any species of earthworm are also influenced by the prevailing environmental conditions, they produce fewer cocoons not only when the soil is too dry, but also when it is too wet (Evans and Guild, 1948). Another factor that can affect the numbers of cocoons produced is the nutrition of the adults which produce them, for instance, earthworms fed on different food materials such as sewage sludge, straw or farmyard manure produced less than one-tenth of the cocoons produced by worms fed on bullock or horse droppings, or peat (Evans and Guild, 1948).

TABLE 3

Growth periods in weeks of several species

Emerged	*A. caliginosa*	*A. chlorotica*	*A. longa*	*D. subrubicunda*	*A. rosea*	*E. foetida*	*L. castaneus*	*L. rubellus*
November 1944	—	42	—	—	—	74	—	—
December	—	39	38	—	—	—	—	—
January 1945	—	35	—	33	—	70	—	—
February	—	30	—	29	—	74	—	—
March	—	29	—	27	—	—	—	—
April	58	—	71	27	—	—	—	—
May	53	—	70	22	62	—	—	—
June	56	—	64	—	61	—	—	—
July	54	—	58	—	53	47	18	37
August	—	—	54	—	50	—	—	—
September	—	—	51	—	—	—	—	—
February 1946	—	—	40	42	—	—	—	—
March	—	—	—	38	—	—	25	—
April	—	—	—	33	—	—	25	—
May	—	—	—	32	—	—	25	—

(*From Evans and Guild, 1948*)

Temperature also affects the time of development before cocoons hatch, for instance, Gerard (1967) found that cocoons of *A. chlorotica* hatched after 36 days at 20°C, 50 days at 15°C and 112 days at 10°C. The time earthworms take to reach sexual maturity from hatching differs greatly between species and has been investigated for only a few species, mostly in culture. Table 3 provides some data on periods of growth of some species kept in culture (Evans and Guild, 1948). In a field site, Satchell (1967) reported that *L. terrestris* matured in one year but environmental factors greatly affected the period of growth to maturity. *A. chlorotica* took 29–42 weeks to mature at field temperature, those worms that matured during the winter months taking longest, but Gerard (1967) calculated that *A. chlorotica*, hatching in a field site in late July, took only 21 weeks to mature. Graff (1953) recorded that the same species took 17–19 weeks to mature when kept at 15°C, and Michon (1954) recorded that they matured in 13 weeks at 18°C. Michon (1954) has shown that the type of food available also affected the length of the maturation period.

TABLE 4

Time of development for some lumbricid worms

Species	No. of cocoons per worm per year	Incubation time of cocoon (weeks)	Period of growth of worm (weeks)	Total time for development (weeks)
E. foetida	11	11	55	66
D. subrubicunda	42	8½	30	38½
L. rubellus	106	16	37	53
L. castaneus	65	14	24	38
E. rosea (*A. rosea*)	8	17½	55	72½
A. caliginosa	27	19	55	74
A. chlorotica	27	12½	36	48½
A. terrestris f. *longa*	8	10	50	60

(*From Wilcke, 1952*)

The life span of mature lumbricids in the field is probably quite short, often no more than a few months (Satchell, 1967), although he calculated that their potential longevity was 4–8 years (Lakhanis and Satchell, 1970), because they are exposed to many hazards. In protected culture conditions however, individuals of *A. longa* have been kept for $10\frac{1}{4}$ years, *E. foetida* for $4\frac{1}{2}$ years and *L. terrestris* for 6 years (Korschelt, 1914) although in the field they are unlikely to

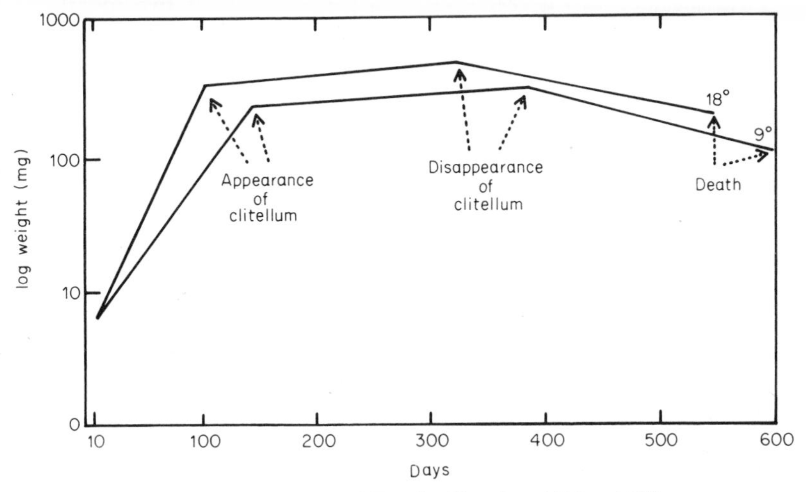

Fig. 19 Development of *D. subrubicunda* at 18°C and 9°C. (*Michon, 1954*)

attain such ages. Michon (1954) studied the life spans of other species and reported that they ranged from 15 months (*A. chlorotica*) to 31 months (*A. longa*), for a group of ten species kept at 18°C, with an average for all species of two years. Data for time of development of some lumbricid worms were given by Wilcke (1952) (Table 4).

Earthworms stop breeding some time before they die. Figure 19 (Michon, 1954) shows that for *D. subrubicunda* the active breeding period, which is when the clitellum is prominent, is only half the adult life span (Michon, 1954).

The life cycles of members of few other families have been studied in any detail, but they probably differ from those of lumbricids only in time. Thus, the cocoons are usually produced when climatic conditions are suitable. Bahl (1922) reported that although *Pheretima* spp. produced cocoons throughout the year in culture in the subtropical climate of India, most were produced from March to June, and few during the rainy months of July and August. Fully clitellate individuals of the acanthodrilid worm *Diplocardia egglestoni* occurred only during May and June in an area of Michigan during a dry summer (Murchie, 1958). The breeding period of the marsh-dwelling worm *Criodrilus lacuum* (Glossoscolecidae) is restricted to a period of about two months, from the end of April to the end of June, and Stephenson (1929) suggested that such a short breeding period may be common among marsh-dwelling oligochaetes. *Alma* sp., a glossoscolecid worm from Egypt, lives in the mud of ditches, and produces cocoons before the mud dries up during the hot season; this enables it to survive adverse conditions. Little is known of other aspects of the life cycles of these less-studied families.

The incubation period of cocoons of *Pheretima hilgendorf* varies from 244 to 264 days, which is considerably longer than that of most lumbricid species. Murchie (1960) calculated that individuals of the species *Bimastos zeteki* do not become sexually active for 12–24 months after hatching or, if they emerged at the beginning of a dry period, they might take as long as 22–26 months. This indicates that the incubation periods of different species of earthworms differ greatly. Grant (1956) thought that *Pheretima hupeiensis* had an annual life cycle.

3.2 Reproduction

3.2.1 *Spermatogenesis*

Groups of spermatogonia formed from follicles (masses of reproductive cells) in the testes, pass into the seminal vesicles, these containing male cells at all stages of development. These cells eventually form spermatozoa which float freely in the seminal vesicles at first, but pass quickly into the testes sacs and attach themselves to the surface of the sperm funnels where they remain until copulation occurs. During mating they are swept by the cilia of the funnel into the vas efferens and hence to the exterior through the male pore via the vas deferens.

3.2.2 *Oogenesis*

The first stages of the development of the ova normally occurs in the basal part of the ovaries, with the formation of oogonia, which divide and form oocytes. These do not divide again but increase in size and accumulate yolk. The oocytes are shed from the ovaries into the ovisacs, where reduction division takes place, by the rupture of the peritoneum. When ripe, the ova are discharged from the oviducts through the female pores, and pass into the future cocoon which is secreted by the clitellum. The ova of eudrilid earthworms are transferred to the ovisacs at a much earlier stage of development, so their ovaries are much smaller, except during the very early stages of sexual development.

3.2.3 *Copulation and fertilization*

Earthworms, although hermaphrodite, are not self-fertilizing, so individuals usually mate and fertilize each other before fertile cocoons are produced, although cocoons are also produced parthenogenetically. *L. terrestris* mates on the surface but other species mate below ground. Most species mate periodically throughout the year, except when conditions are unsuitable or they are aestivating or in diapause.

Methods of copulation are not identical for all species; when individuals of *L. terrestris* mate, two worms which are attracted to each other by glandular secretions, lie with the ventral parts of their bodies together, and their heads pointing in opposite directions (Fig. 20). They come into close contact in the region of the spermathecal openings and where the clitellar region of one worm touches the surface of the other. While copulating, the worms do not respond readily to external stimuli such as touch and light. Large quantities of mucus are secreted so that each worm becomes covered with a slime tube between segment 9 and the posterior border of the clitellum, the two slime tubes adhering to, but remaining independent of each other.

A seminal groove (normally seen as a pigmented line) extends from the male pore to the clitellum, but this groove may not be obvious in some species. Each seminal groove is a depression of the outer body wall formed as a series of pits by the contraction of muscles, the arciform muscles, which lie in the longitudinal muscle

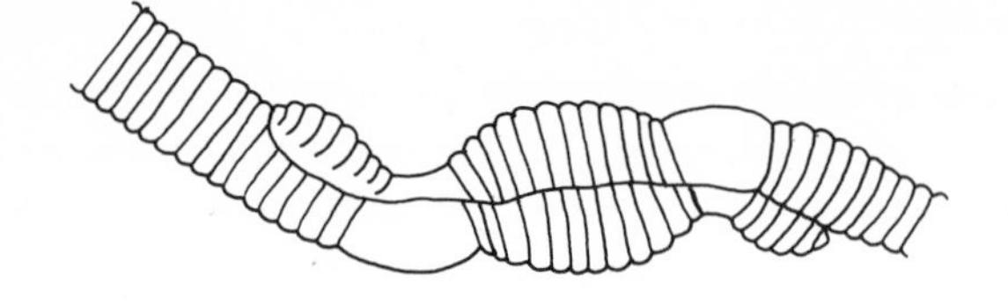

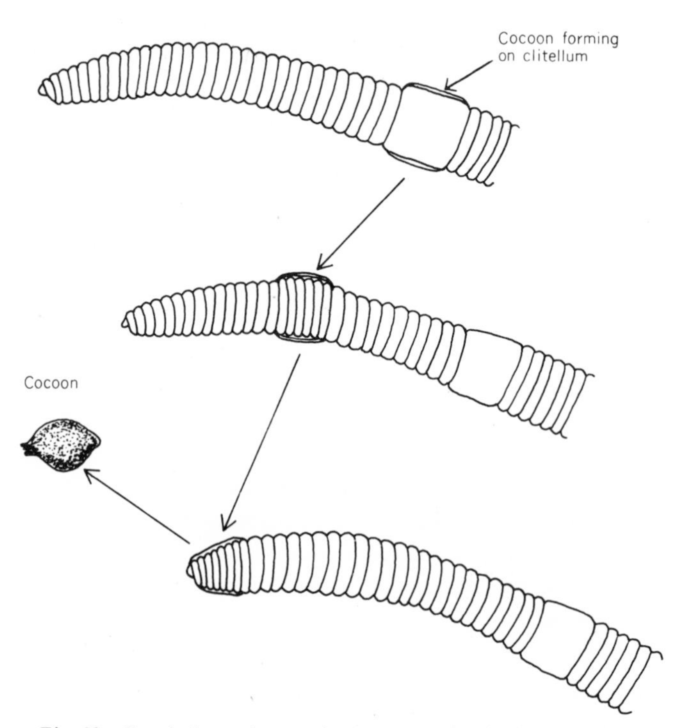

Fig. 20 Copulation and stages in cocoon production in Lumbricidae. (*Modified from Tembe and Dubash, 1963*)

layer. These muscles do not contract simultaneously but successively, beginning at segment 15, each contraction forming a pit. These pits carry seminal fluid as droplets from the male pore to the clitellar region. The seminal fluid collects in the clitellar region, and eventually enters the spermathecae of the opposing worm, although the exact way this occurs is not completely understood. Individuals of *Eisenia foetida* have been seen to clasp and release each other several times, and such movements may assist the entry of the sperm into the spermathecae. Tembe and Dubash (1961) described the copulation of a species of *Pheretima* that has three or four pairs of spermathecae. The male pores first come into contact with the hindmost pair of spermathecal apertures and discharge seminal fluid and prostatic fluid into them. Each worm then moves backwards, and the seminal fluid is discharged into the next pair of spermathecae, until all have been 'charged'. Different methods of sperm transference have been observed in other species, for instance, the octochaetid worm *Entyphoeus waltoni* has spermathecal openings and male pores on raised papillae. The latter, termed penes, are inserted into the spermathecal openings (Bahl, 1927). The eudrilid species, *Schubotziella dunguensis* has a single median male pore formed from two fused copulatory pouches, and a single median spermatheca; during copulation, the two pouches (which communicate with the male ducts) evert, and are inserted into the opening of the spermathecae. All these species that transfer the spermatozoa by direct methods do not form mucus tubes as do lumbricids.

It is important that copulating worms keep close together, and the contours of the ventral surfaces at the points of closest contact help to achieve this. Some setae in segments 6–10 are shorter, thicker, and less curved than normal (Plate 3b). They flex inwards and grip the partner, and the long, pointed and grooved setae on the ventral surface of the clitellar region pierce the body wall of the opposing worm. After copulation, which may take as long as an hour, the worms separate, and each clitellum produces a secretion which eventually hardens over its outer surface. When this is hard, the worm moves backwards, so drawing the tube over its head, and when the worm is completely free, the ends of the tube close, to form the cocoon which is roughly lemon-shaped. The cocoon contains a nutritive albuminous fluid, produced by the clitellar gland cells, the ova, and the spermatozoa which were discharged into it as the tube passed the spermathecal openings. Cocoons continue to be formed until all the stored seminal fluid has been used up. Fertilization is external in the cocoon.

Some eudrilid worms are unusual in that they seem to have a

mechanism for internal fertilization. Sims (1964) described the fertilization of one *Hyperiodrilus* sp. where fertilization is internal in a special chamber (the bursa propulsoria) that is connected to both the spermathecal system and the ovisac.

The cocoon wall consists of interwoven fibrils that are soft when first formed, but later become harder and very resistant to drying and damage. The ends of cocoons are extended into processes or tufts which may be stem-like, conical, or umbrella-like. Cocoons vary in colour from whitish (when formed) to yellow, greenish or brownish, and differ greatly in size. The very large Australian earthworm *Megascolides australis* probably produces the largest cocoons, which may measure up to 75 by 20 mm. However, the size of the earthworm does not always correlate with the size of the cocoon, for example, *L. terrestris* produces cocoons measuring about 6 by 4·5–5 mm, whereas species of *Pheretima* which are about the same size as *L. terrestris* produce much smaller cocoons, about 1·8–2·4 mm by 1·5–2·0 mm. Table 26 gives the sizes of the cocoons of a number of species of earthworms.

The number of fertilized ova in each cocoon ranges from one to twenty for lumbricid worms (Stephenson, 1930) but often only one or two survive and hatch. Out of fourteen species of lumbricid earthworms in culture, only one species (*E. foetida*) commonly produced more than one worm from a cocoon (Evans and Guild, 1948).

Some species can reproduce parthenogenetically, *Allolobophora* and *Lumbricus* species can only breed sexually, but most *Dendrobaena* species are sometimes parthenogenetic and *Octolasium* spp., *Eiseniella* spp., *Allolobophora rosea* and *Dendrobaena rubida f. tenuis* are always parthenogenetic (Satchell, 1967). The male organs of *A. rosea* are sterile, so that production of all cocoons must be parthenogenetic. Asexual reproduction by fragmentation and regeneration occurs in some species of aquatic microdrili but does not seem to appear in true earthworms.

Evans and Guild (1948) investigated whether lumbricid earthworms could produce viable cocoons without mating and fertilization, by rearing isolated individuals of nineteen species in culture. In general, *Allolobophora* sp. produced no cocoons even though they were observed for periods of up to a year. Individuals of *L. castaneus* and *L. rubellus* began producing cocoons five to six months after they became mature, but those of *L. rubellus* were not fertile, and out of a total of 1,704 cocoons produced by *L. castaneus*, only three produced young worms. The other two *Lumbricus* species produced some cocoons, but none were viable. Some individuals of the other species tested produced cocoons at different intervals after they became sexually mature. Most individuals of *D. rubida* and *D. sub-rubicunda* produced viable cocoons soon after they became mature, but a few individuals of the latter species did not produce cocoons for nine months. Most unmated worms of the species *E. foetida* produced cocoons within nine months of maturity, although in the same period mated worms produced more cocoons. Only one individual of *Octolasium cyaneum* and one of *O. lacteum* produced any viable cocoons. The same authors cross-mated individuals of *L. rubellus* and *L. festivus*, which are closely related species, and similar in size. An average of five to six cocoons were produced by each worm, and three months after mating, hybrid worms emerged from some of the cocoons.

3.3 Quiescence, diapause and aestivation

At some times in the year, the soil at the surface may become too dry, too cold, or too warm for earthworms to survive in it. They have several ways of surviving such adverse periods. For instance, the cocoons can resist desiccation and extreme temperatures much better than the worms, and hatch out when conditions once more become favourable. Alternatively, worms may migrate to deeper soil where the moisture and temperature conditions are better. For instance, Madge (1969) observed that two species of eudrilid worms in tropical Africa moved down into deeper soil layers during the dry season. *L. terrestris* can readily move to deeper soil through its more or less permanent vertical burrow system.

Many species of earthworms that normally live near the surface move from the top layers of the soil and become comparatively inactive in deeper soil during adverse periods. There are three states of such inactivity.

1. *Quiescent:* in which the worm responds directly to adverse conditions and becomes active again as soon as conditions become favourable.

2. *Facultative diapause:* which is also caused by adverse environmental conditions, but does not terminate until a certain critical time after conditions become favourable.

3. *Obligatory diapause:* which occurs at a certain time or times each year, independent of current environmental conditions but usually in response to a certain sequence of environmental changes or to some internal mechanism. These stimuli are usually such that adverse conditions tend to occur during the period of diapause.

The term aestivation has been used to cover any or all of these states and it is rarely clear-cut which type of resting phase is occurring. In the temperate zones, earthworms usually become quiescent during warm and dry periods in summer, and also during cold winter spells. Many immature worms of the species *A. chlorotica*, *A. caliginosa* and *A. rosea* become quiescent when the soil is either too dry or too cold (Evans and Guild, 1947), but more earthworms are quiescent during the summer than winter, probably because earthworms can tolerate wet and cold conditions better than hot and dry ones (Gerard, 1967). The difference between quiescence and facultative diapause is probably only one of degree. Gerard (1967) thought that quiescent behaviour enabled worms to exist for long periods on their reserves, when there was little food available in their environment. Lee (1951) stated that in many pasture soils in New Zealand that are subject to severe drought in summer, earthworms move to deeper soil layers and become quiescent. Murchie (1954) reported that *D. egglestoni* can enter a period of quiescence or facultative diapause in response to drought or low soil temperatures. It has been claimed that red-pigmented worms, which include many species of the lumbricid genera *Lumbricus* and *Dendrobaena*, do not go into diapause (Michon, 1960).

At the beginning of diapause, worms stop feeding, empty their guts, and construct a small round or oval cell lined with mucus. They then roll into a tight ball, the body often forming one or two knots in the process, with the two ends of the worm tucked into the centre of the ball; this coiled shape and the mucus-lined cell reduces water loss to a minimum.

A. longa and *A. nocturna* have been reported to go into obligatory diapause (Satchell, 1967). These species, which begin to go into aestivation as early as May, lose all secondary sexual characters such as the clitellum during diapause, and usually come spontaneously out of diapause in September or October. Michon (1954) stated that these species do not diapause if they are kept at 9°C and given adequate amounts of food, but more recent work by Doeksen and van Wingerden (1964) has not confirmed this.

3.4 Growth

According to Hyman (1940), earthworms continue to grow throughout their lives by continually adding segments proliferated from a 'growing zone' just in front of the anus. However, Sun and Pratt (1931) reported that earthworms emerged from the cocoon possessing the full adult number of segments and grew by enlargement of segments, but Gates (1949) challenged these findings. Moment (1953), working on *E. foetida* in culture, stated that newly-emerged worms possessed, on average, the same number of segments as adults, and the only worms that grew by adding segments were those dissected from cocoons. He also observed individual worms from the time they emerged from the egg, and noted that although they increased in size many times, the number of segments remained the same. Probably the correct conclusion as to the way earthworms grow, was given by Evans (1946), who stated that some species possess the adult number of segments on hatching, whereas other species add further segments during post emergence growth. However, species such as *Lumbricus* have fewer segments as they get older even though they continue to increase in size.

L. terrestris in culture in outdoor conditions continued to increase in weight for about three years (Fig. 21) (Satchell, 1967). The increases in weight each year occurred almost entirely during the autumn and spring; during the winter and summer months little weight was gained, and some even lost. After three years the average weight of the worms began to decrease, possibly because some individuals lost weight before dying.

Michon (1954) studied the development and growth of *D. subrubicunda* in culture at two temperatures (Fig. 19). He found that individuals gained weight rapidly until they reached sexual maturity, but after this there was a much slower increase in weight, until the disappearance of the clitellum indicated the onset of senescence.

During this last period there was a slow decline in weight until the death of the worms. Murchie (1960) studied field populations of *B. zeteki* and found that the time taken for individuals to reach full size depended on what time of year they emerged from the cocoons; thus individuals that emerged in autumn and overwintered as small worms, and those that emerged in the spring, grew rapidly, achieving full size by August, to become reproductive the following spring, i.e. within a period of 12–24 months. Those worms that emerged during the summer months, at the beginning of a dry period, grew

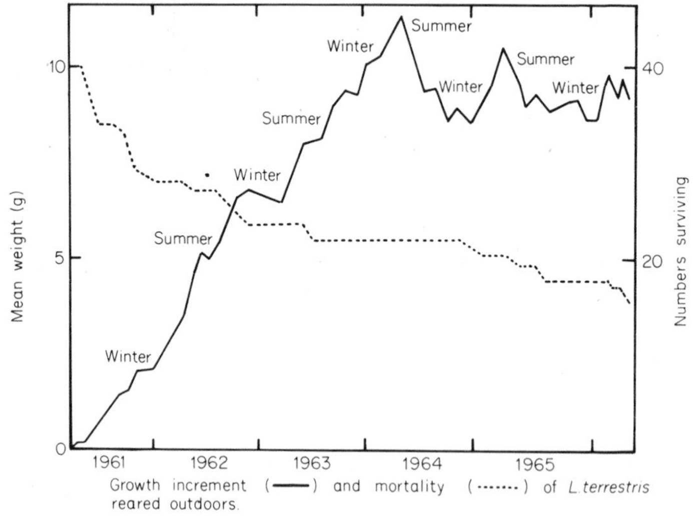

Fig. 21 Growth increment (——) and mortality (------) of *L. terrestris* reared outdoors.
(*After Satchell, 1967*)

less rapidly, and could take 22-26 months to reach maturity. The population of an earthworm species at any one time is made up of young immature, well-grown immature (adolescent), mature and senescent individuals, the proportions depending on the time of year (Fig. 22). The numbers of worms in the first three of these groups for the species *A. nocturna* and *A. caliginosa* were studied by Evans and Guild (1948) in samples taken over a period of nine months from August to May. At the beginning of September the population consisted mainly of immature and adolescent worms but this was followed by a rapid increase in the number of sexually mature worms of both species. The numbers of adolescent worms remained low throughout the rest of the period, with only a small increase in numbers of *A. caliginosa* in spring, although the numbers of adolescent *A. nocturna* increased considerably at this time.

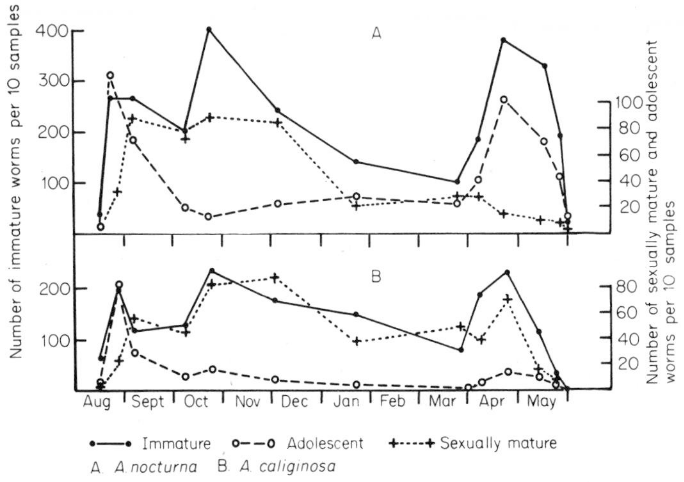

Fig. 22 Seasonal trends in populations of immature, adolescent and sexually mature earthworms.
(*After Evans and Guild, 1948*)

3.5 Behaviour

Many aspects of behaviour have been dealt with already in this chapter and will be discussed further in Chapter 6, but there are some other behavioural patterns that have not been mentioned. All earthworms are very sensitive to touch, the pattern of their reaction varying with both the species and the circumstances. A good example is the behaviour of *L. terrestris* when it scavenges for food on the soil surface. While doing this it usually keeps its tail in its burrow and if it comes into contact with an object such as a stone, it will usually stop and move forward around the object. If the earthworm is touched, it withdraws back into its burrow, sometimes very quickly, and does not emerge again for some time. However, this withdrawal reaction is usually much less violent if the worm is almost completely in its burrow when touched, and it withdraws from the immediate vicinity of the stimulus only. If a worm is grasped while partly out of its burrow it will actively resist an attempt to pull it completely out by extending its posterior setae into the burrow wall and expanding its posterior segments, so as to grip the walls of the burrow and completely fill its exit. It is then very difficult to pull the worm from its burrow, particularly if it is a large individual, and often it will break in two rather than relinquish its grip. After a worm has been subjected to this very violent treatment it usually becomes 'alarmed' and retreats deeply into its burrow, often remaining there for a long period. This behaviour is particularly noticeable when the formalin method of sampling soil for *L. terrestris* is used. If an attempt is made to remove a worm, which has come to the surface under the stimulus of the formalin, before it is clear of its burrow, and the worm is allowed to escape, it does not usually emerge again, even when more formalin is poured into the burrow.

Other species also react vigorously to tactile stimuli, for instance, Murchie (1960) reported that individuals of *B. zeteki* reacted to touch in three different ways. If pricked sharply or handled, they produce a series of lashing movements from side to side, their bodies forming U-shapes with each movement. Certain species of *Pheretima* and *Eudrilus* also do this. Individuals of *B. zeteki* often eject coelomic fluid from the dorsal pores when touched and other earthworm species can eject this fluid to a considerable height, (see Section 1.3). The third and least common response is autotomy, or breaking off posterior segments. Either grasping or impaling the rear end of a worm, may cause it to break at a point a little in front of the stimulated region, and such a break can occur at any point up to thirty-two segments from the posterior end. Stimuli applied to the anterior end does not cause reaction in the same way.

Many other kinds of stimuli applied to earthworms in the soil cause them to come to the surface; these include vibrations caused by a fork inserted into the soil, stamping on the surface, electrical and chemical stimuli.

Individuals of *L. terrestris* feed on leaf and other plant material obtained from the soil surface. They do not feed to any great extent on the leaf material *in situ*, but first pull it into the mouth of the burrow, to a depth of 2·5–7·5 cm, so forming a plug which may

Plate 6a Leaves pulled into mouth of burrows of *L. terrestris*

Plate 6b Mouth of burrows with leaves removed

protrude from the burrow. There have been several suggestions as to why they do this. It may be in order to camouflage the entrance to the burrow, although the plug often makes the burrow more obvious (Plate 6a and b). Alternatively, it may be to prevent the entry of water into the burrow during heavy rain or to keep out cold air. Whatever the reason, these earthworms never leave the mouths of their burrows exposed, and will replug them very quickly if the original plug is removed. If the mouth of the burrow is in an area where there is insufficient organic material to form a plug, they will use inorganic material, commonly small piles of stones termed 'worm cairns' which may be seen in gravel paths, blocking the mouths of burrows. Individuals of *L. terrestris* normally feed on food material only within their burrows, and as far more material is stored than can be used between normal foragings, there is adequate food when inclement conditions on the soil surface prevent them from foraging, or when the local food supply is temporarily exhausted.

L. terrestris carefully selects its food material and pulls most kinds of leaves into the burrows by the tip of the laminae, leaving the non-palatable petioles projecting from the burrow. Darwin (1881) reported that when earthworms were offered paper triangles they always grasped them by the sharpest corner, usually without any attempt to seize any other corner, and this was confirmed by Hanel (1904) although she also found conflicting evidence that if leaves of lime trees were cut so as to round off the pointed apex, worms still grasped them at this point. The general tendency seems to be for the worms to grip the pointed tips of leaves, for instance, Darwin (1881) reported that of 227 leaves of various types, 181 had been drawn in by the tips of the leaves and only 26 had been pulled in by the base of the petioles. In some unexplained way, worms can discern the best way to drag leaves into their burrows with minimum effort. Darwin also described other behavioural patterns shown by individuals of this species during their foraging activities. Baldwin (1917) demonstrated experimentally that *L. terrestris* is much more active when food is available than when it is not.

Many workers have tried to show that earthworms can learn by experience, usually by giving individuals a choice of two courses of action, one of which results in the worm receiving an unpleasant stimulus.

Such experiments usually use a form of T-maze, with the bottom of the 'T' as the entrance, and the left- and right-hand arms of the cross-piece providing alternative routes with stimuli, by either electrodes or a light source in one arm of the 'T', to produce negative or positive responses. The first worker to use this type of apparatus to investigate the ability of earthworms to learn by experience was Yerkes (1912). He used a T-shaped glass box, open at the bottom, resting on a plate-glass base covered with damp blotting paper. The entrance to the bottom arm of the T-maze was joined to a wooden block through which passed a 14-mm-diameter hole lined with blotting paper. The worms were introduced to the maze through this hole. Just inside the entrance to one side arm there was a strip of sandpaper on the glass plate and just beyond, two electrodes, resting on a strip of rubber. (In a previous experiment he used a strip of blotting paper soaked in a strong salt solution instead of the electrodes.) On reaching the junction of the two arms of the cross-piece, the introduced worms could turn either into the arm containing the electrodes which was open at the end, or into the opposite arm, at the far end of which was placed another hollow wooden block, similar to that at the entrance to the T-maze.

On being introduced to the maze, the worms showed no avoidance symptoms, until they entered the arm containing the sandpaper and electrodes. After repeated trials, they learned to avoid contact with the electrodes, and eventually also the sandpaper strip (which acted as a warning of the electrical stimulus), by turning into the opposite arm of the maze. Yerkes also reported that even if up to five anterior segments (including the cerebral ganglia) were removed, individuals still retained the imprinted lesson, but only until the cerebral ganglia were regenerated, thereafter it was lost. Von Heck (1920) also conducted similar experiments using the same type of T-maze, and both Yerkes and he concluded that the worms learned to turn into the arm which did not contain shock-producing electrodes, as a direct consequence of avoidance of the learned negative stimulus in the arm containing the electrodes.

Robinson (1953) tested the reactions of *L. terrestris* instead of *E. foetida* in a similar apparatus. He found that the behaviour of the worms in his maze differed from that observed by Yerkes, in that before reaching the electrode area, they also showed avoidance symptoms to other more generalized stimuli, such as making contact with the walls of the maze. He therefore concluded that Yerkes had over-simplifed the nervous reactions of worms to stimuli and that the process of learning to avoid adverse stimuli was complex. However, Schmidt (1957) compared the behaviour of both species under the same experimental conditions (in a maze similar to that of Yerkes and von Heck), and observed that individuals of *E. foetida* behaved as Yerkes had observed, and those of *L. terrestris* behaved as in Robinson's experiments. He therefore concluded that the behavioural pattern and the ability to learn to avoid an unpleasant stimulus, differed with the species under observation, and the behaviour of one species cannot necessarily be used to predict that of another.

Several other workers (Swartz, 1929; Lauer, 1929; and Bharucha-Reid, 1956) have used similar mazes to study behavioural patterns of worms. Lauer placed electrodes in both arms of the maze, so that the direction of stimulus could be reversed as required. Krivanek (1956) conditioned batches of individuals of *L. terrestris* to avoid high and low intensity light sources, electrical contacts and tactile stimuli. The conditioned state of worms exposed to the high intensity light source was retained for a further period of 48–72 hours after forty-two exposures to the stimulus. This was up to three times as long as that of those worms conditioned to react to the low intensity light source. The conditioned state of worms in the batches exposed to electrical shock and to tactile stimuli was retained for a period similar to that of those exposed to low intensity illumination.

Arbit (1957) also used electrodes in a T-type maze, to examine the behaviour and learning response of two batches of individuals of *L. terrestris* in relation to their diurnal cycle of activity. One batch was tested in the maze between 8 p.m. and midnight, and the other between 8 a.m. and noon. He found that the evening batch needed significantly fewer attempts before they learned to avoid the stimulus, than did the morning batch, and also, that the amount of stimulus needed to start the worms moving in the maze (a camel-hair brush, and shining a light) was significantly less in the evening batch than in the morning one. He concluded that generalizations on learned behavioural patterns are not valid from one species of earthworm to another, and that the time when the pattern was learnt relative to the diurnal cycle is also important. It therefore seems probable that the ability of an earthworm to acquire a learned response to a stimulus depends on its normal activity at the time of learning, because Baldwin (1917) showed that *L. terrestris* is most active between 6 p.m. and midnight.

One phenomenon that occurs occasionally, is that of mass migration of worms for a considerable distance on the surface. Doeksen (1967) reported a number of instances of this in Holland. Individuals of *E. foetida*, which were living in the soil in greenhouses, migrated in large numbers up the sides of buildings and even on to roofs, during damp wet foggy weather. After heavy rain, individuals of the same species living in a dung heap migrated to a nearby farmhouse, climbed up the walls on to the roof, and were even found inside the building. The suggested cause of these migrations was that hydrogen sulphide was produced in the burrows, resulting from anaerobic conditions developing because of poor soil ventilation. Gates (1961) mentioned individuals of *Eutyphoeus* and *Lampito* seen wandering on the surface in daylight after rain in Burma, and that *Pheretima excavatus* and some species of *Dichogaster* were found in trees and on the roofs of high buildings. Mass migrations of *Perionyx* species were also reported by the same author.

Often large numbers of worms appear on the surface of soil after rain and many of these die, probably due to exposure to ultra-violet light or radiation. No adequate explanation of this behaviour has yet been offered.

Phosphorescence or luminescence has been reported to be produced by earthworms. This may show as luminescent blotches or spots on the body, or as a luminous trail behind the worm, but neither of these is long-lived, fading in about thirty seconds. It is believed that the luminescence is due to a luminous slime exuded from the anus or prostomium, but an alternative suggestion is that it is due to coelomocytes in the protective slime. The function of the luminescence may be protective because it occurs usually when the worm is irritated, or in response to vibration or some similar stimulus.

4. Physiology

Most of the studies of the physiology of oligochaete worms have involved terrestrial species, and more is known of the physiology of earthworms than many other aspects of their biology. The reader is referred to a comparatively recent text, 'The Physiology of Earthworms' by M. S. Laverack (1963), for a more detailed treatment than is possible in the present work.

4.1 Respiration

Terrestrial earthworms have no specialized respiratory organs. Oxygen and carbon dioxide diffuse through the cuticle and epidermal tissues into the blood, which contains haemoglobin, a respiratory pigment. The only structural specializations are the greatly branched capillary blood vessels embedded in the body wall. In all respiratory systems, the oxygen must first dissolve in an aqueous layer on the respiratory surface; in the earthworm this is the whole body surface, from where it passes into the body by diffusion, but this is not an active process (Krüger, 1952). The cuticle is kept moist by secretions from the mucous glands of the epidermis.

The haemoglobin in the blood plasma can take up oxygen and transport it to other parts of the body, and the haemoglobin of earthworms can absorb and become saturated with oxygen at pressures as low as 19 mm of mercury (atmospheric pressure is 152 mm). It has been suggested that the coelomic fluid is at a pressure as low as 14 mm (Tembe and Dubash, 1961) so that oxygen can reach the inner tissues even when only small amounts are available. Respiration which depends on simple diffusion is inefficient, but the effectiveness of haemoglobin in the plasma as a respiratory pigment and the low internal pressure of the body fluids of the worm enable it to work very satisfactorily.

Rates of respiration are usually proportional to the surface area available for gaseous diffusion. Smaller earthworms have a greater ratio of surface area to body weight, so in terms of oxygen uptake per g of body tissue, they respire more (Raffy, 1930). The rate of respiration for ***Lumbricus terrestris*** was calculated as being between 38·7 and 45·2 mm^{-3} of oxygen per hour per g of body tissue at 10°C, and 31–70 mm^{-3} at 16–17°C (Johnson, 1942), but Doeksen and Couperus (1968) recorded much higher rates than these.

Earthworms can survive long periods in water and continue to respire, their respiratory rate in this medium depending on the partial pressure of the oxygen dissolved in the water (Raffy, 1930). Some species are adapted to indefinite periods in mud or water-logged soils.

The rate of respiration of earthworms is very dependent on the soil temperature. For instance, the amount of oxygen used increased from 25 to 240 mm^3 per individual per 30 minutes when the temperature changed from 9°C to 27°C (Pomerat and Zarrow, 1936). Thus, earthworms in tropical areas respire faster than those in temperate regions because of higher temperatures. There is some good evidence that earthworm respiration acclimatizes to temperature; thus when earthworms maintained at a low temperature were transferred to a higher temperature, they still respired more slowly than individuals that had been kept at a higher temperature previously (Kirberger, 1953; Saroja, 1959). There is experimental evidence of a diurnal rhythm of oxygen consumption, with maximal rates at about 6 a.m. and 7 p.m. (Ralph, 1957; Doeksen and Couperus, 1968). These times do not necessarily coincide with periods of maximal activity, because it has been suggested that earthworms can accumulate oxygen debts (Ralph, 1957), probably forming lactic acid which can be later resynthesized into glycogen. This may also be the mechanism by which earthworms can survive for many hours with no atmospheric oxygen. It has been shown that during anaerobic respiration, lactic acid and other compounds are formed. For instance, intestinal worms form valerionic acid and it has been suggested that earthworms also accumulate this compound (Lesser, 1910; Stephenson, 1930).

Respiration does not usually decrease until the partial pressure of oxygen in the soil falls to a very low level. At 25% of the normal pressure (38 mm mercury), respiration is depressed by 55–60% (Johnson, 1942). Such low levels of oxygen are usually associated with large amounts of carbon dioxide, and earthworms can respire at the normal rate until the soil atmosphere contains up to 50% carbon dioxide. However, earthworms may move away from carbon dioxide concentrations greater than 25%, although such large amounts probably occur only in localized areas in soil (Shiraishi, 1954). It has been suggested that respiration decreases on exposure to ultra-violet light (Merker and Bräunig, 1927).

4.2 Digestion

The digestive system (described in Chapter 1) is a simple one consisting of a buccal chamber, pharynx, oesophagus, crop, gizzard and intestine. Earthworms derive their nutrition from organic matter, in the form of plant material, living protozoa, rotifers, nematodes, bacteria, fungi and other micro-organisms, and decomposing remains of large and small animals. Most of these are extracted from the large quantities of soil that pass through the gut, although some species, such as *L. terrestris*, feed on leaves directly, and even show preference for particular species and conditions of leaves.

The process of digestion has been worked out in detail for one species of earthworm, *Eisenia foetida*, by van Gansen (1962). The forepart of the digestive system (segments 1–14), which he termed the 'reception zone', contains the sensitive, prehensile mouth, the oesophagus and the ductless pharyngeal gland which secretes an acid mucus containing an amylase but probably no proteolytic enzyme. Other workers have reported that a proteolytic enzyme is secreted by the 'salivary' glands of some species. Opening into the oesophagus is the calciferous gland, which secretes amorphous calcium carbonate particles coated with mucus. The function of this secretion is unknown, but it has been suggested that it influences the pH of the intestinal fluid (Robertson, 1936). Van Gansen believed that its function was the regulation of the calcium level of the blood, but it may be that it is merely responsible for excreting excess calcium carbonate taken up from soil.

Segments 15–44 were termed the 'secretory zone' by van Gansen.

This contains the crop, which leads to the gizzard and intestine. The strong, muscular action of the body wall of the gizzard brings opposing surfaces of the thick internal cuticle against one another, thus grinding up the soil and organic matter which then passes on into the intestine. *E. foetida* secretes two proteases and one amylase mainly from the epithelial brush-border cells, the 'goblet' cells of the intestinal wall secreting mostly mucus. Different enzymes have been reported from this zone for other species, for instance, a lipase and a protease with a rennin-like action (Millott, 1944; Arthur, 1965) and a cellulase and chitinase (Tracey, 1951). Most of the available evidence indicates that the cellulase and chitinase are secreted by the gut wall, although it has been suggested that these enzymes might be produced by symbiotic bacteria and protozoa as in other invertebrates. No doubt other species of earthworms produce many other enzymes, in view of the diverse nature of the food; for instance, one worker reported that an invertase was present in the intestine. The digested food passes into the bloodstream through the intestinal epithelium and is carried to the various parts of the body and tissues for use in metabolism or storage.

The last zone (segments 44 to the anus) in *E. foetida* was termed the 'absorption zone'; here the undigested matter in the intestinal contents becomes enveloped by a peritrophic membrane that lines the intestine and when excreted covers the casts. The composition and structure of the casts will be discussed in a later chapter.

4.3 Excretion

The principal excretory organs in earthworms are the nephridia, which extract waste materials from the coelomic fluid in which they lie, and excrete them to the exterior through the nephridiopores, as urine consisting mainly of ammonia and urea (although some workers have reported uric acid and allantoin in this fluid). There are at least two nephridia in each segment (see Chapter 1), but in some species there are many more. In a few genera such as *Pheretima* the nephridia open into the gut.

Considerable amounts of nitrogenous matter are excreted from the body wall as mucus. This acts as a lubricant, binds soil particles together to form the wall of the burrow and also forms a protective coat against noxious materials. About half the total nitrogen excreted per day is in this mucus. There are several other excretory mechanisms in earthworms, but these are still poorly understood and although there have been many hypotheses, their relative importance and precise mode of action still remains obscure. However, it is known that most of the excretory products of metabolism reach the blood or coelomic fluid from where they must be excreted. Special cells, called chloragogen cells, which belong to the coelomic epithelium of the intestine, are believed to be important in removing excretory materials from the blood, although it has been suggested that they also transport nutrients between tissues and organs (van Gansen, 1956, 1957, 1958; Roots, 1957, 1960). Chloragogen cells, which are always closely associated with blood vessels and capillaries, collect yellow refractive granules which have been termed chloragosomes; these contain a complex arrangement of substances (Roots, 1960). It is generally agreed that the chloragogen cells break off from the surface of the alimentary canal, fall into the coelom and form vacuoles which discharge into the coelomic fluid. Other cells completely disintegrate and their contents, believed to be mainly urea and ammonia, possibly with some fats and glycogen, are liberated into the coelomic fluid where they can be eliminated later. It seems that the chloragogen cells function very much like a mobile liver, acting as a homeostatic device to maintain required levels of substances in the blood and coelomic fluid. The substances deposited in the coelomic fluid by chloragogen cells can be eliminated later either directly by the nephridia, or by being taken up by amoebocytes which are then either deposited in the body wall or in bulky nodules in the coelomic fluid. Some earthworms, such as *Pheretima*, have lymph glands where amoebocytes accumulate. Amoebocytes also occur in the blood, some being deposited in the intestinal wall, and later falling off into the intestine to be excreted with the faeces. Other excretory cells termed uric or bacteroidal cells, are peritoneal cells containing rod-like inclusions, believed to be small crystals of uric acid; these function very much like amoebocytes.

The most important excretory organs are probably the nephridia, which have several postulated modes of action. Firstly, coelomic fluid containing excretory materials in solution passes through the nephrostome and along the nephridial tube by ciliary action, the cilia acting as a sieve. It has been suggested that whole chloragogen cells may also pass into the nephridia but it is unlikely that they do so through the nephrostome, and if this does occur, it is more probably through the walls of the nephridia. Some earthworms have nephridia that are closed internally and thus must function differently, probably by waste materials diffusing across the nephridial wall into the lumen.

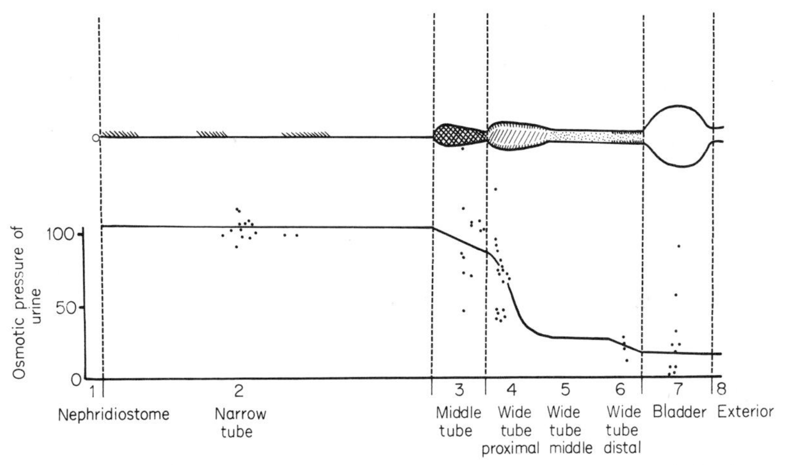

Fig. 23 To show the osmotic pressure of the urine at different levels in the nephridium. The osmotic pressure of the ringer surrounding the nephridium has been equated to 100. Individual observations are shown as points: the line drawn through the points represents the interpretation placed upon them.
(*After Ramsay, 1949*)

The nephridia obviously act as a differential filter, because there is much more urea and ammonia, but less creatinine and protein in the urine they produce, than in the coelomic fluid. The composition of the urine changes as the fluid passes along the nephridium (Ramsay, 1949) (Fig. 23), but it is not known whether osmotic changes are due to resorption of salt or secretion of water. Certainly, the urine is at a much lower osmotic pressure than the coelomic fluid, but how this is achieved is not clear. There is some evidence that granules of waste material can be taken up by the ciliated tube of the middle wall and remain there, so that these parts of the nephridia may act as kidneys of accumulation. Bahl (1947) concluded that nephridia have three functions in excretion, namely filtration, resorption and chemical transformation. He believed protein was reabsorbed through the nephridial wall against a concentration gradient, but this remains to be confirmed and was not accepted by Martin (1957). A possible mode of functioning of nephridia is given diagrammatically in Fig. 24 summarizing some of the excretion and resorption processes.

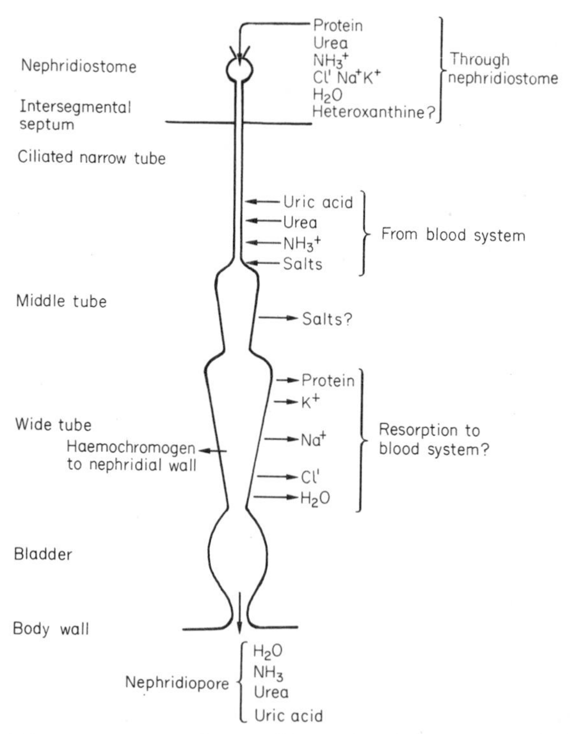

Fig. 24 Diagram to summarize possible mode of functioning of oligochaete nephridia.
(*After Laverack, 1963*)

4.4 Circulation

The structure of the circulatory system of *L. terrestris* (described in Chapter 1) (Figs 7 and 8) is similar to that of most terrestrial oligochaetes. Blood flows forward along the length of the dorsal blood vessel which is the main collecting vessel and which lies in close contact with the gut all along the body. Most of the blood from the dorsal vessel passes down into wide pulsating vessels ('hearts') containing valves, although some blood is also passed anteriorly to the head. The function of the 'hearts' is to pump the blood down into the large ventral vessel in which the blood flows forward anteriorly to the head and backward posteriorly, distributing it to all parts of the body.

Blood is collected from anterior tissues and organs by branches of the subneural vessel (which lies close to the nerve cord) and it flows backwards in this vessel, being eventually returned to the dorsal vessel by the dorso-subneural vessels which run in the septa of each segment. The ventral vessel supplies blood to various organs of the body. To supply the nervous system, branches pass from the ventral vessel to ventro-parietal vessels, then into the lateral-neural vessels, which run alongside the nerve cord and branch into capillaries that supply blood to the cord. From the nerve cord, blood drains into the subneural vessel via branched capillaries.

The oesophagus and intestine receive blood from the ventral vessel through ventro-oesophageal and ventro-intestinal vessels. Blood returns from the oesophagus in the front of the body through paired lateral-oesophageal vessels which lead to the dorso-subneural vessel of the 12th segment and thence to the dorsal vessel. In the posterior part of the body the blood passes from the peri-enteric plexus of capillaries around the intestine back to the dorsal vessel via the paired dorso-intestinal vessels. Three small vessels in each segment run from the typhlosole into the dorsal vessel.

In each segment behind the 12th, blood passes from the ventral vessel to the ventro-parietal vessels, which have numerous branches that end in capillaries close to the skin; these allow the blood to become oxygenated. Blood returns from the skin via other capillaries which lead to the dorso-subneural vessels and thence to the dorsal vessel.

4.5 Nervous system

4.5.1 *General*

The most important part of the nervous system is the large ventral nerve cord (Figs 13 and 14), which contains three longitudinal giant fibres lying in its dorsal part; one median and two smaller ones situated latero-ventrally. The detailed functions of the components of the nervous system are still not fully understood, although they have been extensively studied. Nor is much known about the functioning of the sense organs, synapses, motor fibres and innervation of muscles.

The various sense organs which lie in the body wall are associated with numerous intra-epidermal nerve fibres that end freely between the epidermal cells. It has been suggested (Smallwood, 1926) that the sense organ fibres synapse with the sub-epidermal network of fibres, which are collected together to form the segmental nerves, and the stimuli from them are then passed on to the motor nerves and fibres. Sensory stimuli in one segment may invoke a motor response either via fibres in the nerve of the same side of the same segment, the contralateral nerve of that segment, the nerve before or after that stimulated in the same segment, or even the segment in front or behind. Nerve connections extend for as many as three segments backwards or forwards via association neurones. Stimuli from one segment can also reach other segments indirectly, because the sensory fields of the nerves supplying the body wall, cover more than one segment.

Strong stimuli can by-pass the association neurones and be transmitted up and down the giant fibres. Giant fibres are not completely continuous along the length of the body because in each segment there is an oblique partition across each fibre which is probably not a true synapse, but clearly is no obstacle to rapid transmission of stimuli. This arrangement of giant fibres is important in allowing the transmission of motor impulses along the nerve cord at high speed (600 metres per second) and a very rapid reaction of the animal to adverse stimuli. There is some evidence that the median, dorsal, longitudinal giant fibre transmits stimuli only from the front to the back of the worm and not vice versa. There are also smaller lateral giant fibres and it is believed that these transmit stimuli in the opposite direction. The probable explanation is that most of the sensory input from the front of the animal passes into the median giant fibre and in the rear of the animal into the lateral fibres.

Most of the available evidence indicates that the main chemical transmitter substances in the central nervous system are acetylcholine, probably adrenalin, noradrenalin and possibly 5HT. The main function of the enlarged cerebral ganglia in the front of the body seems to be mostly inhibitory, because if they are excised, the worm moves continuously. They seem to play little part in initiating movement.

It is now well established that certain nerve cells are capable of elaborating and releasing complex organic substances which act as hormones. These can be released into the bloodstream and act upon some distant organ. Such neurosecretions were first reported for *L. terrestris* by Schmid (1947), who found that secretions were produced cyclically and then production could be initiated by adrenalin or novocaine.

Secretory cells have been reported from the cerebral, suboesophageal and the first two ventral ganglia (Herlant-Meewis, 1956; Hubl, 1956; Marapo, 1959). It seems possible that bodily functions such as reproduction, pigment migration, development of secondary sex characters, activity cycles and diurnal respiratory cycles may be controlled by neurosecretions, but little evidence of this is available, except in a few instances. It has been shown that the yearly reproductive cycle of *L. terrestris* seems to be controlled by neurosecretions, because immature individuals lack certain secretory cells in the cerebral ganglia (Hubl, 1953). If these ganglia are removed from mature worms, the secondary sex characters disappear and egg laying and cocoon production ceases. It has also been demonstrated that neurosecretions are important in regenerative phenomena.

4.5.2 *Light reactions*

Earthworms do not have recognizable eyes, but do have sensory cells with a lens-like structure in regions of the epidermis and dermis, particularly on the prostomium. The middle part of the body is rather less sensitive to light and the posterior is also slightly sensitive. Earthworms certainly respond to light stimuli, particularly if they are suddenly exposed to it after being kept in the dark. *L. terrestris* is photopositive to very weak sources of light and photonegative to strong ones, i.e. they crawl towards dim lights and away from strong ones. However, if they are kept for long periods in strong light they do not react at all to a sudden increase in intensity

(Hess, 1924); it has been suggested that this is due to saturation of the light receptors. Other species such as *Pheretima* spp. are completely photonegative and respond in proportion to the intensity of the light (Howell, 1939).

Earthworms react differently to different wavelengths of light; blue light is stimulating and red is not, so the activities of experimental earthworms are best studied in red light. Ultra-violet light seems to be harmful, and it has been suggested that some of the earthworms lying dead on the surface of the ground after rain have been killed by the ultra-violet light of the sun (Merker and Bräunig, 1927). The mode of action of the light-sensitive reaction is not yet clear, but it has been shown that some drugs depress photosensitivity and cutting the ventral nerve cord also modifies reactions to light. Small electric potentials have been reported in the nerves from light stimulation (Prosser, 1935). All the evidence is that the reaction to light is controlled and coordinated by the cerebral ganglia. Howell (1939) suggested the following mechanism for the photonegative response in *Pheretima agrestis*: impulses from photic stimuli to light-sensitive cells travel along different nerves to the cerebral ganglia, the ventral nerve cord and the circumpharyngeal connectives. Strong impulses cross over directly in the transverse commissure of the cerebral ganglia and in the commissures of the ventral nerve cord. Weak impulses are relayed to the cerebral ganglia where, with impulses entering these ganglia from the cerebral nerves, or from the circumpharyngeal connectives, they are probably modified and cross over in the transverse commissure. After crossing over, the impulses go to the muscles of the side opposite that which was illuminated, causing them to contract, thus producing a negative response.

4.5.3 *Chemoreception*

Darwin (1881) showed that earthworms could readily distinguish between different food substances and it has been shown more recently that *E. foetida* can react to diverse chemical stimuli. The functions of such reactions are probably to help in selection of food, give warning of adverse environmental conditions such as soil acidity, and assist in mating by detecting the mucous secretions of other earthworms.

The sense organs, which react to chemical stimuli, are on the prostomium or the buccal epithelium which comes into contact with substances when the buccal chamber is everted during feeding. Several workers have demonstrated the selection of and preference for particular forms of leaf litter by earthworms (Mangold, 1951; Wittich, 1953). Satchell (1967) showed that the palatability of leaves to earthworms depended greatly on their polyphenol content. Mangold (1953) reported that alkaloid substances above a certain concentration were not accepted by earthworms. Earthworms reacted inconsistently to glucose and saccharose (Mangold, 1953). It seems from this evidence that most food preferences of earthworms depend on chemoreception.

Earthworms can also detect acids. Mangold (1953) showed that acids such as phosphoric, tartaric, citric, oxalic and malic found in plant materials were accepted at low concentrations but not at high ones. Different species differ in their tolerance to soil acidity, but all have a threshold of pH below which they cannot live for long, so the ability to detect pH is essential for survival. Laverack (1961) demonstrated that *Allolobophora longa* will not burrow into soil with a pH below 4·5 and *L. terrestris* into soil with a pH below 4·1. He showed that acid-sensitive fibres are present all over the body by taking recordings of nerve impulses which occurred when worms were dipped in buffer solutions. There was no indication that the prostomium was more sensitive than the rest of the body. The threshold values for the acid-sensitive organs were pH 4·4 to 4·6 for *A. longa*, 4·1 to 4·3 for *L. terrestris* and 3·8 for *Lumbricus rubellus*, and these organs reacted only to acid stimuli and not to sodium chloride, quinine or sucrose. Clearly, earthworms are sensitive to chemical substances other than those named so far, because they are rapidly brought to the soil surface when dilute solutions of potassium permanganate or formaldehyde are poured on to the soil.

4.5.4 *Thigmotactic reactions*

Earthworms are very sensitive to touch because of tactile receptors on discrete areas on the surface of the body. There are three nerves in each segment involved in touch, the anterior of these having a greater receptive area in the segment ahead, and the posterior one a greater sensory area on the segment behind. These tactile organs are involved in the thigmotactic responses of earthworms. When placed on soil or other surfaces, worms become very active until they find a suitable crevice or crack, so that the sides of the body are in contact with the substrate, when presumably the touch receptors are stimulated and the worm stops moving. This response is very strong and can overcome the light reaction.

4.5.5 *Response to electrical stimuli*

Earthworms respond rapidly to electrical stimuli, becoming U-shaped with both ends directed towards the cathode when a voltage is applied. If the direction of the current is reversed, the worm reorientates itself (Moore, 1923). Worms also emerge on to the surface of soil to which an electric current is applied (Satchell, 1955).

4.6 Water relationships

Conservation of water to avoid dehydration is very important for all terrestrial animals. Earthworms have a thin, permeable cuticle overlying a mucus-secreting epidermis, neither of which can do much to prevent water losses, so that terrestrial species are exposed to considerable risks from dehydration due to water losses through the skin. Water can also escape from the body via the mouth, anus, dorsal pores and nephridia.

Up to 85% of the fresh weight of earthworms is water, a considerable part of this being in the coelomic fluid and blood; so they must be able to prevent excessive water losses. Earthworms in soil are often not fully hydrated and they may increase in weight by as much as 15% if placed in water (Wolf, 1940) and lose this when replaced in soil. Thus, earthworms must have mechanisms for replacing or conserving water, although their locomotion and burrowing, which depend to some extent upon hydrostatic pressure, are not seriously affected by losses of water up to 18%. There are diurnal changes in weight of 2–3% but these are of little significance (Wolf, 1940). Amounts of moisture in soil vary greatly, and if they fall too low, earthworms eventually begin to lose moisture and weight, although as much as 70–75% of the body weight can be lost without killing the worm (Roots, 1956). As the worm becomes more and more dehydrated so its behaviour changes; first there is a reaction termed a 'dehydration tropism' (Parker and Parshley, 1911) which is governed by the prostomium, because if this part of the body is removed, the reaction does not occur. There is then a period when the body rolls in the soil and eventually coelomic fluid is expelled from the dorsal pores in an attempt to moisten the surface of the body. Ultimately, rigor, anabiosis and death occur. If irreversible changes have not occurred, the animals can be revived by submersion in water even after very large water losses.

There has been disagreement as to the ways in which earthworms maintain a reasonably constant concentration in the internal fluid. Greatly different estimates of the osmotic pressure of the coelomic fluid have been reported, although it is clear that the osmotic pressure of the blood of *Lumbricus* is slightly below, and that of *Pheretima* above, that of the coelomic fluid (Ramsay, 1949).

Probably most of the uptake and loss of water is through the body wall. There is evidence (Stephenson, 1945) that *Lumbricus* can maintain a constant internal salt concentration, because if these worms are placed in dilute salt solutions, the internal chloride concentration remains above the external concentration. Worms can also keep their internal salt concentration at a level lower than that of very concentrated solutions in which they are placed. This was confirmed by Ramsay (1949), who showed that as the concentration of chloride in the medium increased, so the osmotic pressures of the body fluids also increased, keeping them always greater than those of the medium. Urine excreted from the nephridiopores is always at a lower concentration (hypotonic) to the body fluids, except in very concentrated media with a sodium chloride content greater than 1%. Thus, although earthworms can maintain a relatively constant internal osmotic pressure in dilute solutions, they are unable to do so in concentrated ones. Although, it has been clearly shown that earthworms maintain their internal fluid at a relatively constant concentration, and excrete hypotonic urine as do freshwater animals, the exact mechanisms of this process are still not clear.

Some oligochaetes are aquatic, and terrestrial earthworms differ greatly in their affinity for and ability to survive in flooded soil, ranging from semi-aquatic species to those that prefer dry soil. Nevertheless, earthworms of the species *Allolobophora chlorotica, A. longa, Dendrobaena subrubicunda, L. rubellus* and *L. terrestris* which normally inhabit dry soils, were all able to survive from 31 to 50 weeks in soil totally submerged beneath aerated water. Even then, the factor limiting survival was probably lack of food rather than submersion. Specimens of *A. chlorotica* have made burrows in flooded soil (Roots, 1956) and cocoons of this species will hatch under water.

D. subrubicunda, *L. rubellus* and *L. terrestris* show preference for moist, rather than saturated or flooded soil, but the other species discussed above, sometimes choose waterlogged soil, although in general, most terrestrial species leave flooded soil for drier sites and worms are often found on the soil surface after rain. There is evidence that individuals of some species become acclimatized to water.

Obviously the osmotic regulation and excretory systems of earthworms are such that in water they can excrete fluids as fast as they diffuse inwards, and so maintain internal fluids at a constant pressure.

4.7 Locomotion and peristalsis

When earthworms burrow through the soil, they do so by co-ordinated contractions of the longitudinal and circular muscle bands that lie in the body wall. These contractions are made possible by the segments being kept turgid by the coelomic fluid, although there is no movement of fluid across the septa between segments when peristaltic waves pass along the body. If coelomic fluid moved along the body it would be difficult for one end of a worm to contract and the other to expand.

The prostomium is used to find a cavity suitable for burrowing. The earthworm protrudes the setae on its posterior segments to keep the rear end of the body fixed, then contracts the circular muscles of the front end of the body, thus causing the anterior segments to extend forwards with a thrust of between two and eight grams according to the size of the worm (Fig. 25). This thrust depends mainly on the internal hydrostatic pressure of the worm, and the area to which the pressure is applied. The contraction of the circular muscles passes from segment to segment backwards along the body, and after the wave of contraction has passed, the anterior longitudinal muscles contract in turn, drawing the posterior end of the body forwards. The rear end of the body is then held again

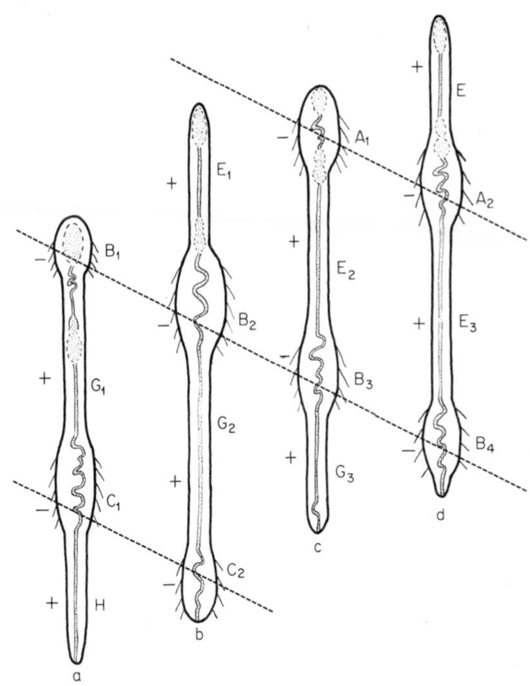

Fig. 25 Patterns of muscular configuration of an earthworm, with concomitant pattern of the intestine and possibly of the oesophagus. The wavy line represents the gut in the folded condition:

+ represents regions of high pressure; − represents regions of low pressure.

(*After Arthur, 1965*)

in position in the burrow by the backward-pointing setae, and another wave of contraction passes backwards down the body. The setae are protracted and retracted as required, by special muscles; when the movement of the body is reversed, so is the direction of the setae.

Segmental contractions are by no means completely under control of the nerve cord, although if the cord is cut, the body muscles posterior to the cut no longer contract. The segmental contractions seem to be caused by intersegmental stimuli, because if an earthworm is cut in half, but the two parts of the worm joined by thread and then put under tension, peristaltic waves continue to pass down the body.

Nevertheless, if the body of the animal is cut through completely except for the nerve cord, contractions can still pass down the body but much faster (about 25 mm per second). During normal peristalsis the nerve cord has an electrical rhythm whose frequency is identical with that of the muscular rhythm (Gray and Lissman, 1938). Thus, although locomotion can be maintained either by nervous impulses along the nerve cord or by reflexes passed from one segment to the next, normally the two methods reinforce each other. The cerebral ganglia seem to play no part in locomotion.

Contractions can be stimulated by touch, as well as by longitudinal tensions ranging from 0·1 to 1·0 g; and they can be blocked by anaesthesis with magnesium chloride. It is believed that the tension is detected by receptor cells lying between the ventral and dorsal setae of each side in every segment, and by other cells which lie buried in the circular muscle layer. If the tension is removed, peristalsis stops unless the ventral surface of the earthworm is in contact with the substratum. Peristalsis can also be initiated in a decapitated earthworm, by making the head electropositive to the posterior end, and inhibited by reversing the flow of the current (Gray and Lissman, 1938). Thus, the ways in which peristalsis is maintained are now understood, but we still do not know the ways in which it is initiated.

The gut also has peristaltic contractions which seem to be mainly under nervous control. Two sets of nerves reach the gut, one from the circumoesophageal nerve ring forming a plexus between the mucous membranes and the muscle layers of the intestine, and another from the ventral nerve cord running up each septum to the gut. The former nerves are believed to be excitatory and the latter inhibitory. However, it still is not clear how the gut maintains its peristaltic movements.

4.8 Regeneration

Earthworms have very great powers of regenerating injured or lost parts of their bodies. There has been a very considerable literature on the morphogenetic aspects of this phenomenon, but very little on its physiology or mechanism. It is not proposed to deal with the detailed histology of the regenerating tissues here, the reader is referred to Stephenson (1930) for a full treatment of this subject.

Oligochaetes can regenerate either the anterior or posterior portions of their bodies, but the posterior part grows again more readily than the anterior. When the front end is replaced, wound tissue is formed over the cut end, taking about seven days to grow, the gut being either occluded or remaining open during regrowth. The rear end of the body usually remains open during regeneration and the wound tissue forms a ring. When the front of the body regenerates, it grows with the same width as the rest of the body, but the rear usually grows as a slender appendage, which develops all the segments that it can regenerate before it begins to expand in width. Regenerated tissue takes about two to three months before becoming fully pigmented.

Regeneration of the front end of the body does not occur if more than a limited number of segments are cut off, and even the posterior end will not grow again if too many segments are removed, e.g. anterior fragments of *E. foetida*, consisting of less than thirteen to thirty segments regenerate only slowly, if at all. The full number of segments is not always regenerated; the number regrown depending on the length of the remaining fragment of the body. There is a gradual decrease in regenerative capacity along the body from front to back. Moment (1953) showed that if an individual of any species was bisected at any particular segment, there was a remarkable consistency in the number of segments regenerated. No earthworm ever continues regenerating segments until there are more than the uninjured animal possesses. Moment believed that growth was controlled by electrical means. Earthworms have a voltage difference from one end to the other, and this is the same in young worms as in old ones. When a worm is cut in two, this voltage decreases sharply, but returns to its original level within three weeks. Moment (1953) believed that the cessation of growth after a certain number of segments have been produced, could be explained on the basis that each segment has a particular voltage contribution, and when the full number of segments is attained, the total critical voltage inhibits further growth. There is some evidence to support such a theory, but none to show the mechanism of such an electrical inhibition of growth.

Some species have less ability to replace lost segments than others, e.g. *E. foetida* regenerates segments more readily than does *L. terrestris*. The other feature essential for regeneration to occur seems to be that the nervous system in the part remaining should be intact. If the anterior end of an earthworm is cut off and the nerve cord removed from a few of the remaining segments, these segments are reabsorbed and regeneration starts at the segment that still has its nerve cord (Carter, 1940). If the nerve cord is removed completely, no regeneration occurs (Zhinkin, 1936). Nerve depressants such as lithium, acetylcholine, parathion and disulfoton also greatly inhibit regeneration. Oxygen is necessary for rapid regeneration and no regrowth occurs under anaerobic conditions; too much oxygen can also block regeneration. This has been confirmed by treatment with potassium cyanide which changes both the rate of respiration and also that of regeneration.

Some evidence exists that the chloragogen cells are important in regeneration, because there is a mass migration of these cells to the wound, after part of the worm is cut off. There is also evidence that temperature influences regeneration, all species regenerating much more readily in summer, with optimum temperatures between 18°C and 20°C, much higher than those which normally favour development of terrestrial earthworms. Younger worms regenerate segments much more readily and quickly than do older ones. The sexual organs are rarely regenerated if the part of the anterior end which carries them is amputated.

Sometimes abnormal regeneration occurs, for instance, a posterior end may grow in place of the head, or vice versa. If the anterior part of individuals of *E. foetida* is removed by a cut made behind the groove between segments 18/19, either an abnormal head or a tail is grown.

Thus, although the phenomenon of regeneration has attracted considerable attention, there still exists no adequate knowledge of its mechanism.

4.9 Transplantation

Some workers have experimented on transplanting portions of earthworms on to other individuals (Tembe and Dubash, 1963). When the anterior portion of one earthworm is sutured to the tail portion of another of the same species in the normal position, the intestine, blood vessels and nerve cords become continuous within two weeks. The parts also unite satisfactorily, even if one half is rotated 90° to the other. If two tails are joined they remain alive for a considerable time, but if two heads are joined they do not usually unite satisfactorily, and the parts do not survive for long.

Ovaries have been satisfactorily transplanted from one worm to another, even when the ovaries of an individual of *L. terrestris* were transplanted into another worm of the species *Allolobophora caliginosa.*

5. Ecology

5.1 Estimation of populations

5.1.1 *Handsorting*

To estimate earthworm populations, some method of determining the number of worms in small sample areas is necessary. Most of the early studies involved digging up soil samples and sorting these by hand (Stockli, 1928), and indeed many workers still use this method, except that they take cores or quadrats of soil of exact dimensions to enable accurate population estimates to be made. Workers who have estimated populations by handsorting include Bretscher (1896), Bornebusch (1930), Ford (1935), Hopp (1947), Reynoldson (1955), Low (1955), Svendsen (1955), Wilcke (1955), Barley (1959), Van Rhee and Nathans (1961), El-Duweini and Ghabbour (1965). Zicsi (1962) compared the efficiency of samples of sizes 0·06, 0·25, 0·5 and 1·0 m^2, taken with a square sampling tool, for estimating populations of earthworms by handsorting. He concluded that sixteen sample units of an area of $\frac{1}{16}$ m^2 taken to a depth of 20 cm gave an adequate estimation of a population of medium-sized species. For larger worms and deeper samples, a larger area of sample was required.

Nelson and Satchell (1962) tested how many earthworms could be recovered from soil by handsorting, by introducing known numbers of worms into soil. They found that smaller worms and dark-coloured worms were often missed, and their numbers under-estimated; when 924 worms were introduced to soil, 93% of all worms were recovered by handsorting, but only 80% of immature *Allolobophora chlorotica* and 74% *Lumbricus castaneus* were found. They concluded that handsorting was satisfactory only for individuals of more than 0·2 g live weight. There were considerable differences in efficiency between individual sorters.

5.1.2 *Soil washing*

Morris (1922) and Ladell (1936), used a method of washing soil with a jet of water through a series of sieves. Raw (1960b) handsorted samples from rather poor pasture soils, and then washed the soil away from the same samples in a 2 mm-mesh sieve within another 0·5 mm-mesh sieve standing in a bowl of water. The sieves were then immersed in magnesium sulphate solution of specific gravity 1·2 and the worms that floated to the surface collected. Only 52% (84% of the weight) of the worms collected were found by handsorting, and a further 48% were recovered by washing. From a heavy, poorly-structured arable soil 59% (90% by weight), and from a light, well-drained soil 89% (95% by weight) of the total numbers found were recovered by handsorting. Obviously washing is more efficient than handsorting, and it also recovers cocoons, but the washing method takes much longer. A mechanized soil washing method, which involves rotating the containers in which the sieves stand (Edwards *et al*, 1970), is much faster and suitable for most soils.

5.1.3 *Electrical methods*

Fishermen have obtained earthworms from soil for bait by attaching one lead of an a.c. mains to a copper wire attached to a non-conducting handle and inserting it into the soil. Walton and Johnstone-Wallace (1937) reported that such a technique could be used for sampling earthworm populations, and Doeksen (1950), who experimented further found that a steel rod 8–10 mm in diameter and 75 cm long with an insulated handle, was suitable as an electrode. The voltage he used was 220–240 V at 3–5 A, and the strength of current was regulated either by a variable resistance, or by inserting the electrode deeper into the soil. One to three electrodes could be used simultaneously.

The conductivity of the soil depends on its moisture content, but usually the current penetrates deep into the soil, bringing worms up from deep burrows; however, if the surface soil is dry it may drive the worms downwards instead. This possibility can be minimized by insulating all of the electrode except its point. Worms usually emerge between 20 cm and 1 m from the electrodes, but there is some danger that worms close to the electrode may be killed by the current. Nevertheless this method does seem effective in sampling for deep-living worms.

Satchell (1955) also used an electrical method with a 2KVA generator that led to a water-cooled electrode inserted 46 cm into the soil, with a voltage of 360 V applied. The portable generator made the method more suitable for field sampling. Satchell believed the main defect of the method was that of defining the exact limits of the volume of soil from which earthworms were recovered.

5.1.4 *Chemical methods*

The first chemical extractant used to sample earthworm populations was mercuric chloride solution (1·7–2·3 litres of solution containing 15 cc $HgCl_2$ in 18·25 litres water) (Eaton and Chandler, 1942). Evans and Guild (1947) used potassium permanganate solution to bring worms to the soil surface (1·5 g per litre at a rate of 6·8 litres per m^2), and later used this method in their population studies (Guild, 1948; Evans and Guild, 1947; Guild, 1952). Jefferson (1955) used Mowrah meal (the material remaining after oil is extracted from ground seeds of the Bassia tree, *Bassia longifolia*). Raw (1959) reported that a 0·55% formalin solution (25 ml of 40% formalin in 4·56 litres water applied to 0·36 m^2 of soil surface) was very effective in bringing earthworms to the surface. This is because dilute formalin is less toxic to worms than potassium permanganate, which often kills worms before they reach the surface. The main disadvantage of these chemical methods is that they do not recover all species equally efficiently; those species with wide burrows coming to the surface much more readily than the non-burrowing species. Satchell (1969) used the same method but recommended that much more solution (three applications of 9 litres of 0·165–0·55% per 0·5 m^2) should be used. He pointed out that the soil temperature and soil moisture content both affect the number of worms coming to the soil surface and worked out a correction factor based on a regression analysis, which would correct for the soil temperature at the time of sampling (Satchell, 1963), and he later modified this (Lakhani and Satchell, 1970). This, however,

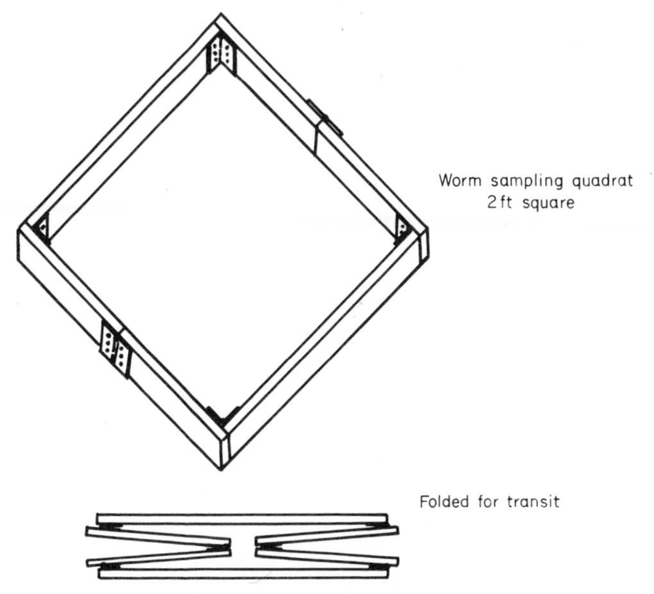

Fig. 26 Quadrat for formalin sampling.

could not correct for worms that were aestivating and did not respond to the chemical. A suitable quadrat for formalin sampling is shown in Fig. 26.

5.1.5 *Heat extraction*

This method, which is little used but may be useful in obtaining small surface-living species from matted turf, involves a container (55 × 45 cm) with a wire sieve 5 cm from its bottom. Soil samples (20 × 20 × 10 cm deep) are placed on the sieve, immersed in water with fourteen 60 W light bulbs suspended above, and left for 3 hours after which worms can be collected from the bottom of the container.

5.1.6 *Comparisons of methods*

Several workers have compared the relative efficiency of extracting earthworms from soil by two or more of these methods. Svendsen (1955) reported that handsorting was much more efficient than using potassium permanganate. Raw (1959) compared formalin with potassium permanganate, and from one arable orchard he obtained 59·7 worms from 0·36 m^2 with formalin, 32·5 with potassium permanganate and 47·5 by handsorting. Comparable figures for a grass orchard were 165·1 for formalin, 83·9 for potassium permanganate, and 280·0 for handsorting. Bouché (1969) compared handsorting with first applying formalin, then handsorting to find animals that had not been extracted. He reported that 55·4 worms per m^2 were extracted by formalin and a further 273·4 per m^2 by handsorting soil from the same area.

There seems little doubt that handsorting or washing give the best results for most species but are very time consuming. At present, the formalin method seems the best compromise for species with burrows.

5.1.7 *Number and size of samples*

Many different sizes of samples have been used, ranging from soil cores 20 cm diameter to 50 × 50 cm and 1 m^2 quadrats. Zicsi (1958) reported that the number of worms he recovered per m^2 by handsorting decreased with increasing size of sample. The minimum sample area required depends very much on the density of a population in a site. For most purposes, half or quarter m^2 quadrats seem suitable.

5.2 Size of populations

Earthworm populations can be expressed either in terms of numbers or weight (biomass). Numbers are sometimes misleading because they do not differentiate between very small and large individuals. It is probably best to express populations both as numbers and biomass, and most workers on earthworms do this.

It is often difficult to obtain the live weight of large numbers of earthworms directly, and Satchell (1969) described a method for calculating the live weight of earthworms that had been kept in 10% formalin solution. He plotted a regression of the live weights of worms against their weight after being in 10% formalin, then oven-dried them at 105°C, and reweighed them. He obtained the expression: 1 g dry weight = 5·5 g live weight. The gut contents of an earthworm may be as much as 20% of its total live weight so this must be accounted for when estimates of population biomass are made.

Few workers have sampled a variety of habitats at the same time

TABLE 5

Numbers and weights of earthworms in different habitats

	No/m²	g/m²	Site	Extraction method	References
Fallow soil	18·5–33·5	4·6–8·4	U.S.S.R	Handsorting	Dzangaliev and Belousova (1969)
Pseudotsuga mor	14·0	4·7	N. Wales	Handsorting	Reynoldson (1955)
Pine woodland	40	17	Hants., U.K.	Formalin	Edwards and Heath (unpublished)
Oak woodland	184	68	Hants., U.K.	Formalin	Edwards and Heath (unpublished)
Mixed woodland	68	37	Herts., U.K.	Formalin	Edwards and Heath (1963)
Mixed woodland	157	40	N. Wales	Handsorting	Reynoldson (1955)
Arable land	146	50	N. Wales	Handsorting	Reynoldson (1955)
Arable land	287	76	Bardsey Island	Handsorting	Reynoldson *et al* (1955)
Arable land	220	48	Germany	Wet sieving	Krüger (1952)
Arable land	18	1·6	Herts., U.K.	Formalin	Raw and Lofty (unpublished)
Arable land with dung	79	39·9	Herts., U.K.	Formalin	Raw and Lofty (unpublished)
Orchard (grass)	848	287	Cambs., U.K.	Formalin and Handsorting	Raw (1959)
Orchard (grass)	300–500	75–122	Holland	Handsorting	van Rhee and Nathans (1961)
Orchard (grass)	254–344	63·5–86·0	U.S.S.R	Handsorting	Dzangaliev and Belousova (1969)
Pasture	389–470	52–110	Westmorland, U.K.	Handsorting	Svendsen (1957)
Pasture	390	56	Bardsey Island	Handsorting	Reynoldson *et al* (1955)
Pasture	481–524	112–120	N. Wales	Handsorting	Reynoldson (1955)
Pasture	260–640	51–152	N.S.W. Australia	Handsorting	Barley (1959)
Under pig litter	960	272	U.S.A.	Formalin	Edwards (unpublished)

of year, so it is difficult to assess how different habitats influence earthworm populations. Some typical populations estimated by handsorting or formalin extraction are given in Table 5. These estimates are very approximate because of large variations in efficiency of extraction, and seasonal changes in numbers of earthworms. For instance, populations range from less than one to 850 per m^2 (0·5–300 g per m^2). There are fewer earthworms in mor soils, fallow soils and moorlands, than in mull soils, with the numbers in regularly cultivated arable soils very variable, and intermediate in size between these and pasture. The populations in mull woodlands, orchards and pastures are rather similar.

There is little information on populations in tropical soils. Block and Banage (1968) recorded populations of between 7·4 and 101·8 per m^2 (0·23–3·64 g per m^2) in Uganda soils, and Madge (1969) 33 per m^2 (10 g per m^2) in Nigerian grassland. Duweini and Ghabbour (1965) reported populations ranging from 8 to 788 per m^2 (1·96–578·2 g per m^2) in Egyptian soils.

5.3 Population structure

Not much information is available about the structure of populations of earthworms. There are usually more immature than adult worms in a population, although their relative proportions vary seasonally, for instance, Raw (1962) reported that the proportions of *Lumbricus terrestris* individuals of different ages in his samples from an orchard were in the ratio eight mature worms to thirteen large immatures and thirty-one small immatures.

Evans and Guild (1948) sampled twelve fields at Rothamsted for earthworms and compared the number of immature and adult worms of different species (Table 6). There is no doubt that these proportions are not a fixed ratio, but that they depend on the time of year when a population is sampled, so that during active breeding periods the proportion of immature worms will be greatest. There is some indication however, that the structure of populations of different species may differ. Figure 27 shows the seasonal changes in proportions of adult and immature worms of *Allolobophora nocturna* and *Allolobophora caliginosa* found by van Rhee (1966), and Fig. 28 those of *A. chlorotica*; clearly at all times of the year the numbers of adults exceed those of immatures. Van Rhee (1967)

TABLE 6

Ratios of immature worms to adults during October

Field	*A. nocturna*	*A. caliginosa*	*A. chlorotica*	*A. rosea*	*L. terrestris*
Parklands	4·5	3·0	0·5	2·2	10·3
Great Field I	1·9	1·4	0·5	1·9	12·3
Great Field III	2·5	1·5	0·3	1·8	11·1
Pastures	3·1	0·7	0·2	1·2	7·3
Longcroft	2·3	1·6	0·6	3·4	6·9
Claycroft	1·5	1·1	—	1·8	7·5
Appletree	5·3	1·0	0·5	2·1	5·0
Great Field II	4·5	1·2	0·3	2·5	19·3
Parklands Wedge	3·5	1·5	—	5·0	12·9
New Zealand	2·0	1·7	0·3	1·7	—
Stackyard	12·8	2·3	0·6	3·0	5·7
Delharding	3·3	2·5	1·6	1·9	1·5

(*From Evans and Guild, 1948*)

compared the population structure of *L. terrestris*, *L. castaneus*, *Allolobophora rosea*, *A. caliginosa* and *A. chlorotica* in four orchard soils in five successive years, and for all species except *L. castaneus*, the immatures greatly outnumbered the adults. For the species *Pheretima hupeiensis*, in north-west United States, mature individuals are abundant in August but die after reproduction, and

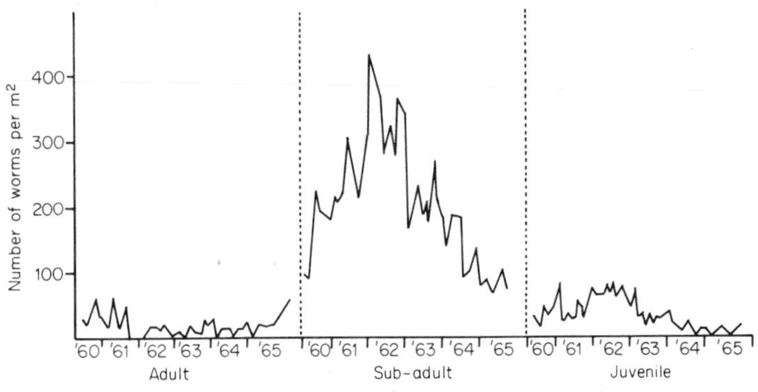

Fig. 27 Age-class composition of *A. caliginosa*. (*After Van Rhee, 1966*)

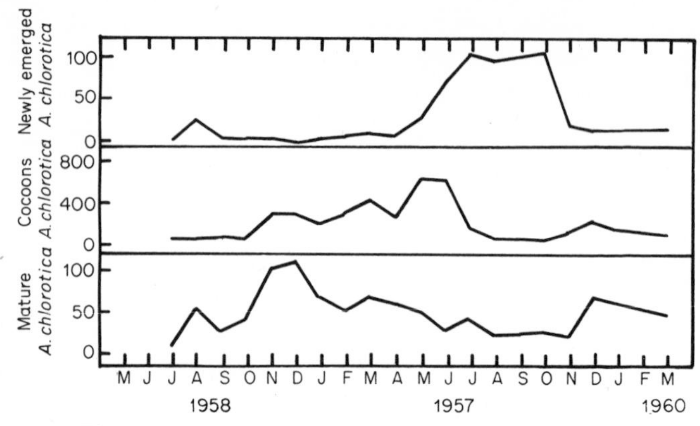

Fig. 28 Seasonal abundance of *A. chlorotica* in 'Pastures' at ROTHAMSTED 1958–60 (*Gerard 1960*)

immatures are numerous in September, but by November, the entire population has become immature and remains so throughout the winter months (Grant, 1956).

5.4 Population distributions

5.4.1 *Horizontal distributions*

Earthworms are by no means randomly distributed in soil (Guild, 1952), and Murchie (1958) classified possible factors responsible for variability in horizontal distributions as:

(1) physico-chemical (soil temperature, moisture, pH, inorganic salts, aeration and texture);
(2) available food (herbage, leaf litter, dung, consolidated organic matter);
(3) reproductive potential and dispersive powers of the species.

Murchie concluded that no single one of these factors was likely to be solely responsible for distribution, but rather the interaction of several factors.

There is not much experimental evidence that physico-chemical factors other than soil moisture cause aggregations, but several workers have correlated aggregations with the available food supplies. For instance, in pasture, *Dendrobaena octaedra* and *Lumbricus rubellus* were significantly aggregated beneath dung-pats in spring (Boyd, 1957, 1958) and pigmented species such as *Lumbricus festivus*, *L. rubellus*, *L. castaneus*, *Dendrobaena rubida*, *D. octaedra* and *Bimastos eiseni* were shown to aggregate under dung-pats more than non-pigmented *Allolobophora* species (Svendsen, 1957). If they react so readily to dung, then it is likely that variations in distribution of other forms of organic matter may also influence their distribution. However, aggregations cannot always be explained on the basis of heterogeneity of the habitat; Satchell (1955) showed that *L. castaneus* and *A. rosea* were greatly aggregated in a relatively uniform pasture which had not been grazed. It was suggested that aggregations might occur when earthworms are reproducing more rapidly than the offspring can disperse.

When all species of worms in a habitat are considered, a pattern of overlapping aggregations is usually found. When Satchell (1955) calculated indices of dispersion for adults and immatures, he found that the adults were nearly randomly dispersed but the immatures were aggregated. Hence, a species with distinct seasonal abundance can be expected to pass from a very aggregated phase in the breeding season in early summer to an almost random phase in winter (Satchell, 1955).

Hamblyn and Dingwall (1945) claimed that the rate of advance of the margin of populations of *A. caliginosa* from inoculation points in recently limed grasslands was of the order of 10 m per year, and they suggested that any rapid horizontal dispersal was probably by cocoons carried in soil on agricultural implements, hooves of animals or in streams.

5.4.2 *Vertical distributions*

Different species of lumbricids inhabit different depth zones in the soil (Fig. 29), but the vertical distribution of each species changes considerably with the time of year. The seasonal vertical distribution of the common British lumbricids has been studied by several workers. Species such as *D. octaedra* and *B. eiseni* live in the surface organic horizon of soil for most of the year. *A. caliginosa*, *A. chlorotica*, *A. rosea*, *L. castaneus* and *L. rubellus* commonly occur within 8 cm of the soil surface, as do immature individuals of *Octolasium lacteum*, *Octolasium cyaneum*, *Allolobophora longa*, *A.*

nocturna and *L. terrestris*. Most adult and nearly mature individuals of *O. cyaneum* are in the top 15 cm and although they have distinct burrows, these are usually temporary. *A. longa* and *A. nocturna* have fairly permanent burrows which usually penetrate as deep as about 45 cm, but the deeper vertical burrows of *L. terrestris* commonly go down to a depth of 1 m and can penetrate as deep as 2·5 m. Usually those species that feed on or near the surface are dark-coloured, whereas the subterranean species are predominantly pale.

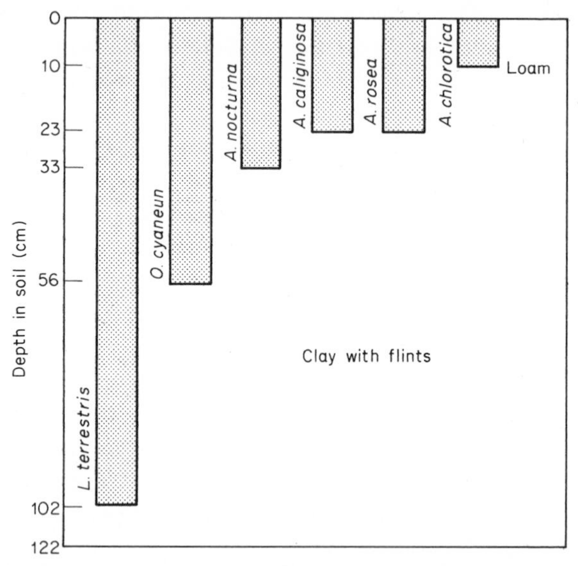

Fig. 29 Vertical distributions of earthworms in a Rothamsted pasture. (*Adapted from Satchell, 1953*)

Gerard (1967) studied the vertical distribution of common earthworms at different times of the year in England (Fig. 30). Most worms in his samples were below 7·5 cm deep in January and February when the soil temperature was about 0°C, but by March, when the soil temperature had risen to 5°C at a depth of 10 cm, most individuals of *A. chlorotica*, *A. caliginosa* and *A. rosea*, and small and medium-sized individuals of *A. longa*, *A. nocturna* and *L. terrestris*, had moved in to the top 7·5 cm of soil, although the larger worms were still deeper in the soil. From June to October, worms of

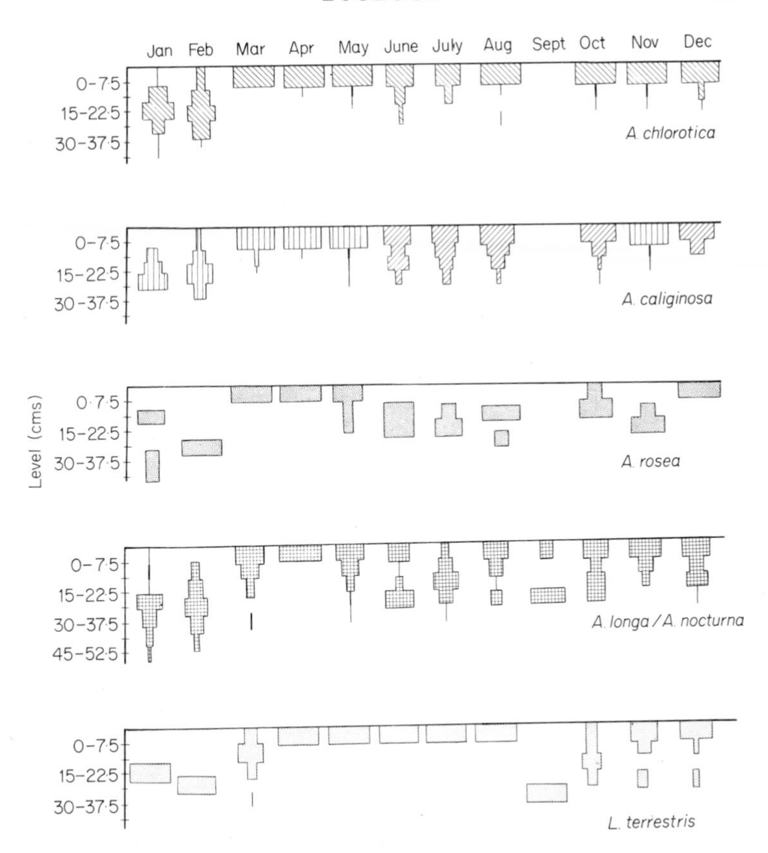

Fig. 30 The depth of six species of earthworms in monthly soil samples from January to December, 1959 (expressed as percentages for each species in each sample). Samples were taken in 3-in layers to a depth of 12 in, and sometimes deeper (up to 21 in).
(*After Gerrard, 1967*)

most species were below the top 7·5 cm again, except for newly hatched individuals. In November, December and the following April, most worms were again in the top 7·5 cm. The two factors influencing movement to deeper soil, seemed to be very cold or very dry surface soil. All species of worms (except *L. terrestris*) seemed to be quiescent in summer and mid-winter, and at both these times were deeper than 7·5 cm below the surface. More worms were quiescent in summer than in winter. Nearly all cocoons were found in the top 15 cm of soil, most being in the top 7·5 cm. Earthworms of the genus *Diplocardia* in South-east United States move from a depth of about 10 cm in October to about 40 cm in January and return to the surface in spring, with a critical temperature for downward movement of approximately 6°C (Dowdy, 1944). In North-west United States, worms of the species *P. hupeiensis* were active in the 15–20 cm soil level and were active near the surface only in September and March, and from November to February they were deeper than 55 cm (Grant, 1956). It seems that in most parts of the world, seasonal vertical migrations of earthworms occur, and these are initiated mostly by the upper soil levels becoming unsuitable for earthworms to feed and grow satisfactorily.

5.5 Seasonal populations and activity

Numbers of earthworms and their activity vary greatly during the annual cycle, but not all workers differentiate adequately between seasonal changes in numbers and activity. This is particularly so of those workers who used chemical extractant methods for determin-

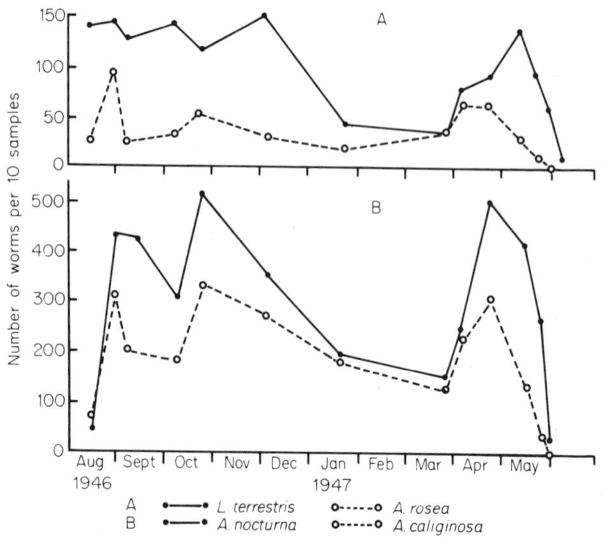

Fig. 31 Seasonal trends in total populations. (*After Evans and Guild, 1948*)

ing populations; such techniques depend on the activity of the earthworms and quiescent individuals do not respond.

Evans and Guild (1947) followed changes in numbers of earthworms in an old pasture field in England for more than a year, using a chemical sampling method (Fig. 31) and they concluded that the two soil conditions which affected activity most were temperature and moisture (Fig. 32), although another factor was the obligatory

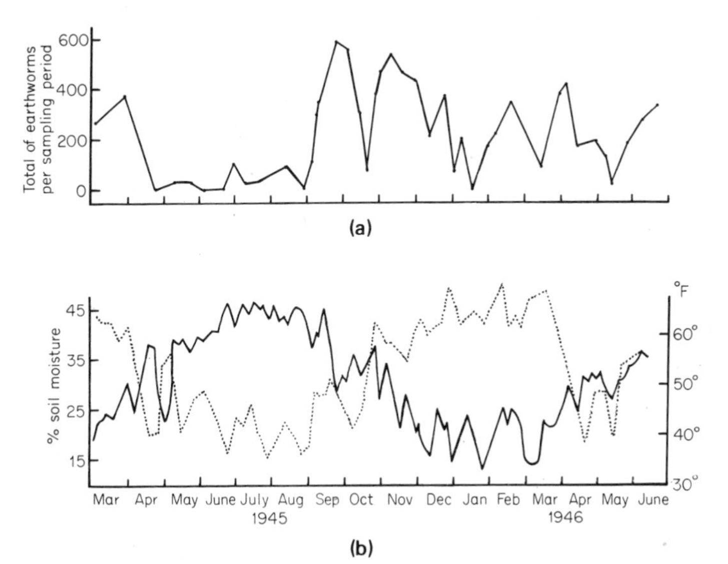

Fig. 32 (a) The seasonal changes in activity of all earthworms (b) the corresponding changes in soil moisture at 2 in, and temperature at 4 in, from March 1945 to June 1946 ...... percentage soil moisture at 2 in; — soil temperature °F. at 4 in. – 0 – total earthworms per ten samples.
(*Evans and Guild, 1947*)

diapause from May to October of the two species *A. nocturna* and *A. longa* or the periods of quiescence or facultative diapause of *A. chlorotica, A. caliginosa* and *A. rosea* during adverse conditions.

Seasonal activity of surface-casting species can also be assessed by the numbers of worm casts produced. Evans and Guild (1947c) found that the numbers of casts deposited, and the numbers of *A. longa* and *A. nocturna* obtained by potassium permanganate sampling were closely correlated (Fig. 33). Other workers have estimated seasonal activity by counting the numbers of worms on the surface at night, but this is very inaccurate because many species of worms only come to the surface when it is wet. Both moisture and temperature influence the activity of earthworms, for instance, Kollmannsperger (1955) estimated that 10·5°C was an optimum temperature for maximum activity.

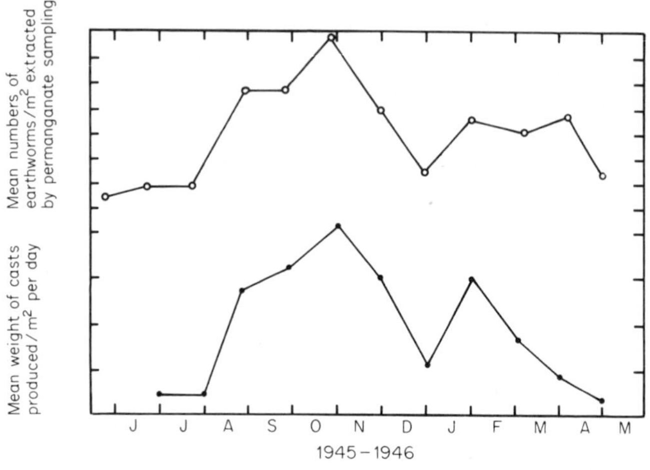

Fig. 33 Seasonal variations in wormcast production and numbers of earthworms extracted by permanganate sampling, Great field Rothamsted.
(*Evans and Guild, 1947*)

Gerard (1967) reported that in pasture soil in England, *A. chlorotica*, *A. caliginosa* and *A. rosea* were usually within 10 cm of the soil surface, but when the soil temperature fell below 5°C, or the soil became very dry, individuals of these species moved to deeper soil. In hot, dry periods in summer, most species became inactive, and again were deeper in the soil. Hopp (1947) believed that the most important factor influencing seasonal changes in numbers of worms in arable soils in the United States, was that worms were killed when the unprotected surface soil became frozen in winter. Seasonal changes in earthworm populations have also been ascribed to other causes, for instance, Waters (1955) suggested that flushes in availability of dead root material and herbage debris were a main cause of increased numbers of worms, but this is not generally accepted.

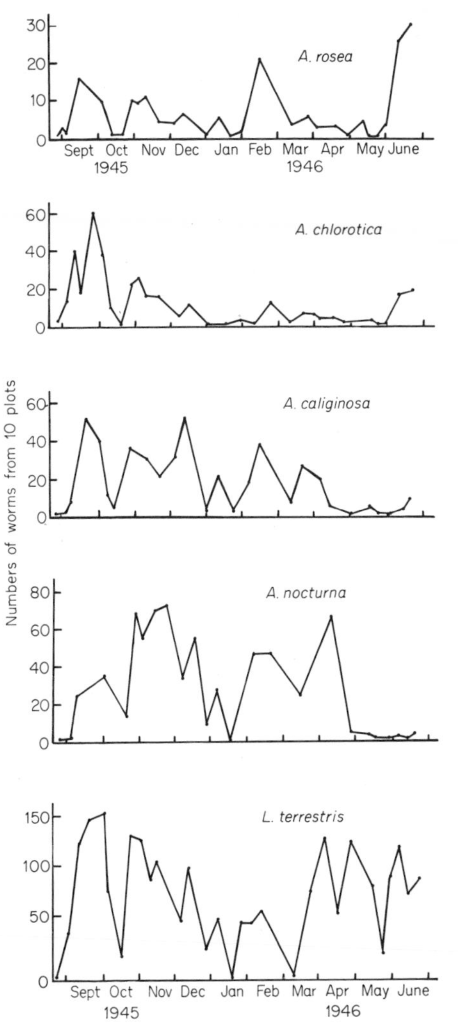

Fig. 34 The seasonal changes in activity of five species of earthworms. (*After Evans and Guild, 1947*)

The five species *L. terrestris*, *A. rosea*, *A. chlorotica*, *A. nocturna* and *A. caliginosa* were most active in English pasture between August and December, and April to May (Evans and Guild, 1947) (Fig. 34). The principal activity of *Hyperiodrilus africanus* in Nigeria was during May and June at the beginning of the wet season,

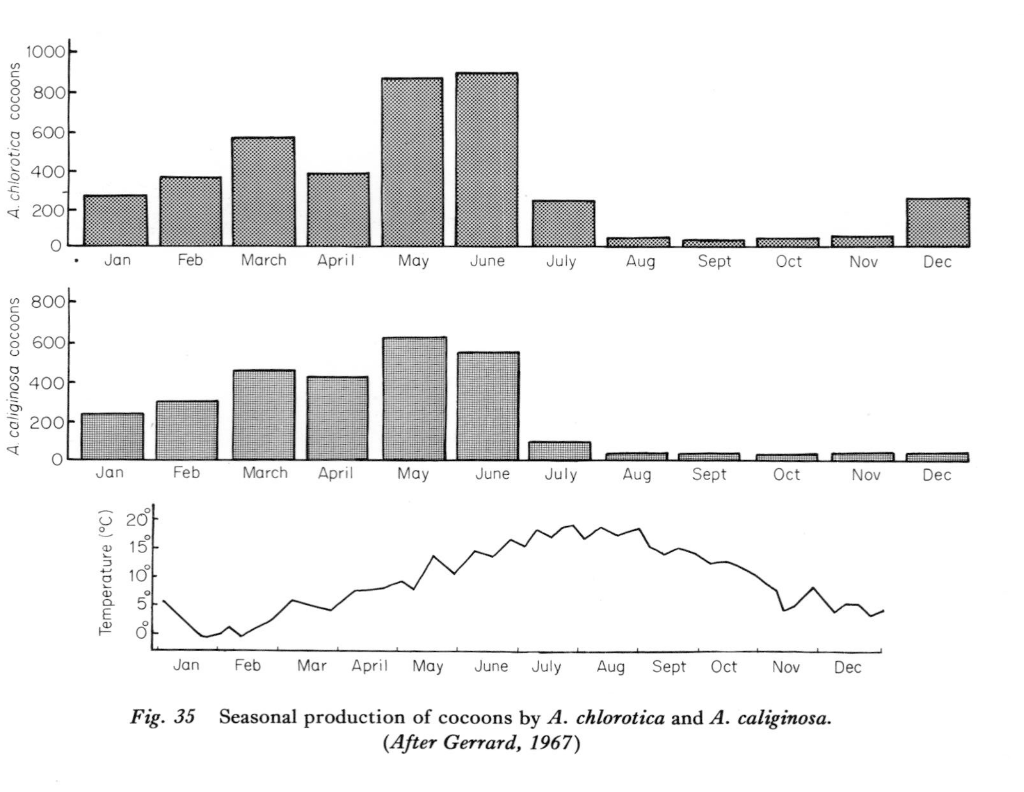

Fig. 35 **Seasonal production of cocoons by *A. chlorotica* and *A. caliginosa*.**
(*After Gerrard, 1967*)

thereafter numbers lessening gradually until November, when few were found, (Madge, 1969). The numbers of cocoons produced also varies seasonally, and Gerard (1967) showed that in an English pasture, most were produced in late spring and early summer (Fig. 35).

Gates (1961) reported that earthworm activity in the tropics is also limited to certain seasons; in the monsoon tropical climate of Burma and the humid subtropical climate of India, earthworms are active mainly in the four to six months of the rainy season between May and October. By contrast, in the humid continental climate of eastern United States they are most active in the spring and autumn months (Fig. 36). Hopp (1947) gave data on seasonal population

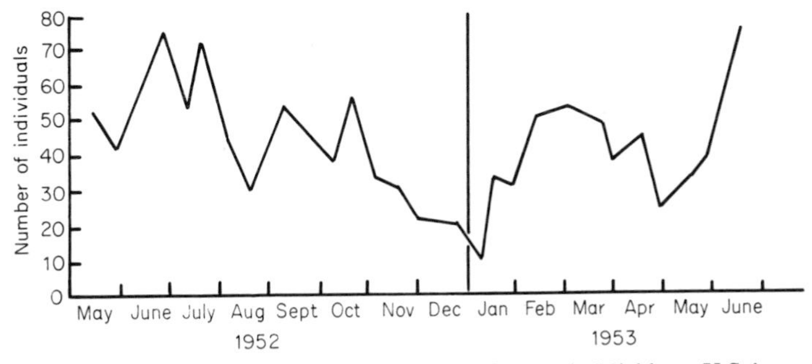

Fig. 36 Seasonal variation in activity of *A. rosea* in Michigan, U.S.A. (*After Murchie, 1958*)

changes in Maryland, U.S.A., which confirmed this, but since he only took samples to a depth of 17·5 cm, the proportion of the total population he extracted varied considerably with the time of year. Grant (1956) reported that *P. hupeiensis* was most active in north-west United States during the summer months, and retreated to soil below 55 cm deep from November to February.

In grassland in Japan, the greatest numbers of earthworms occurred in autumn, especially in October, and numbers were very low in winter, especially during January and February (Fig. 37) (Nakamura, 1968). In Australia, populations of *A. rosea*, *A. caliginosa* and *O. cyaneum* in pasture increased from May to July and decreased from July to October. Soil temperatures were not low

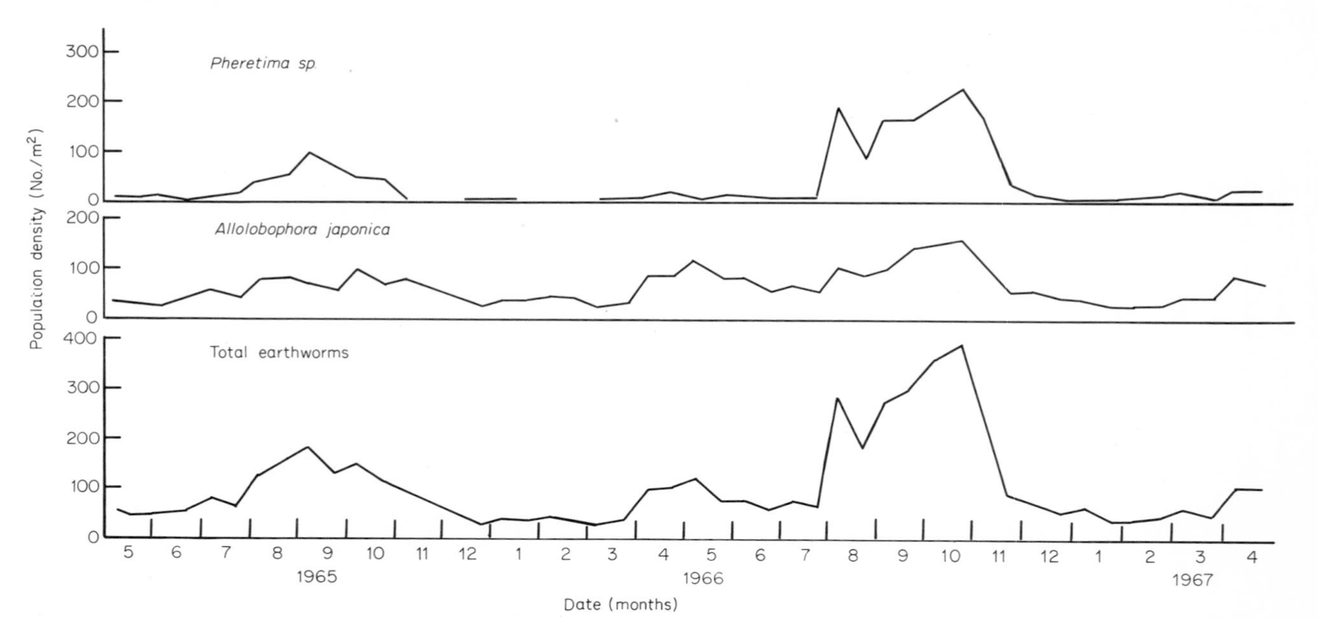

Fig. 37 Seasonal variations in the population density of earthworms in alluvial soil grassland in Japan during the period May 1965 to April 1967.
(*After Nakamura, 1968*)

enough during the wet season (10°C at 15 cm depth) to prevent breeding. The vertical distribution of the worms also changed during the year, so that in winter most were in the top 15 cm and few were below 30 cm, whereas when soils began to dry out in spring, few worms remained in the top 15 cm. During summer, 60% of all worms were between 15 and 30 cm deep. In a New Zealand pasture,

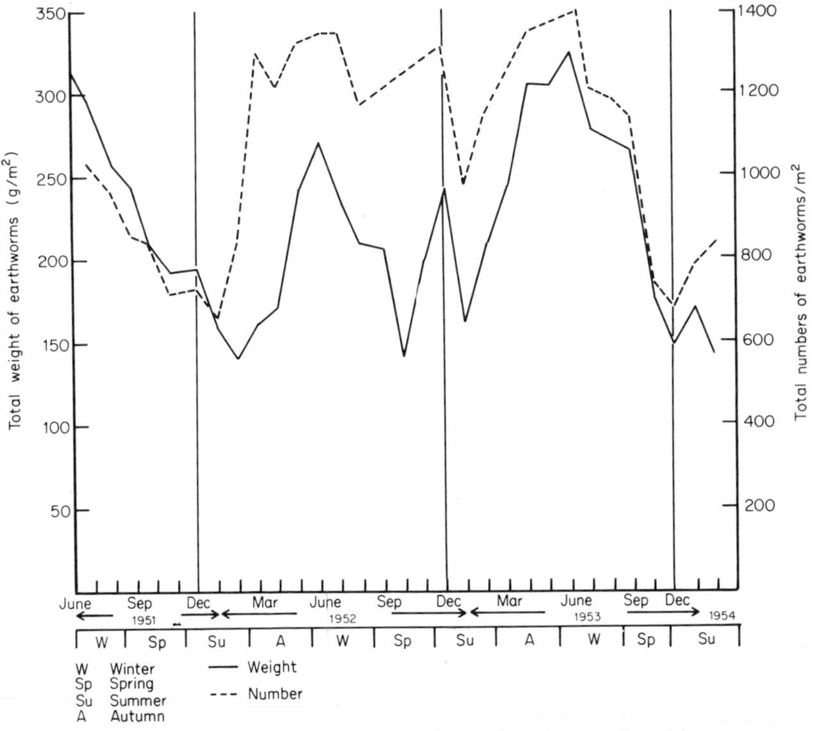

Fig. 38 Seasonal fluctuations in the abundance of earthworms found in pastureland at Palmerston North, New Zealand.
(*Waters, 1955*)

peak populations occurred in mid-winter between 1951 and 1954 (Waters, 1955), (Barley, 1959) (Fig. 38).

5.6 Burrowing and casting

Earthworms form burrows by literally eating their way through the soil and pushing through crevices. Not all species have burrows; it is usually only those species that penetrate deep into the soil such as *L. terrestris*, *A. longa* and *A. nocturna* that have permanent burrows, with smooth walls cemented together with mucous secretions and ejected soil, pressed into the soil interspaces. Often, the mucous secretions serve as a substratum for growth of fungi.

Earthworms can be induced to form burrows in soil between two sheets of glass (Evans, 1947). In such experiments, *A. caliginosa* formed an extensive burrow network through the top 20 cm of soil in a few days, whereas *L. terrestris* took four to six weeks to do this. The speed of burrowing depended very much on the texture of the soil, deep-burrowing species taking four to five times as long to burrow in clay as the same species in light loam. Garner (1953) described a method of making latex casts of the burrows of earthworms by pouring liquid latex (thinned with ammonia and diluted 1 to 8 with distilled water) into the burrows.

Earthworms of the genus *Lumbricus* do not burrow extensively, so long as an adequate food supply is present on the surface, but when food is scarce, burrowing activity is greatly stimulated (Evans, 1947). Burrows of some species such as *L. terrestris* and *A. nocturna*, go down to a depth of 150–240 cm, being vertical for most of their depth, but often branching extensively near the surface. Burrows range from about 3 mm to 12 mm in diameter, but it is not certain whether worms increase the size of their burrows as they grow, or make new ones. Some earthworms can burrow very deep into the soil; *Drawida grandis* has burrows to a depth of 2·7–3·0 m (Bahl, 1950). Most shallow-working species, which are usually smaller, do not have well-defined burrows, although some of the surface species, e.g. *P. hupeiensis* have complex burrow systems in the top 7·5–15·0 cm of soil (Grant, 1956).

It is usually the burrowing species that produce casts on the soil surface near the exits to their burrows. Of the eight to ten common field species of lumbricids in England, only three, *L. terrestris*, *A. longa* and *A. nocturna*, particularly the latter two species, produce casts on the surface of the soil. Usually, they cast more on heavy soils than on light open ones, because in the latter, much of the faeces are passed into spaces and crevices below the soil surface. The numbers of casts produced varies seasonally (Fig. 39) and is a good index of activities.

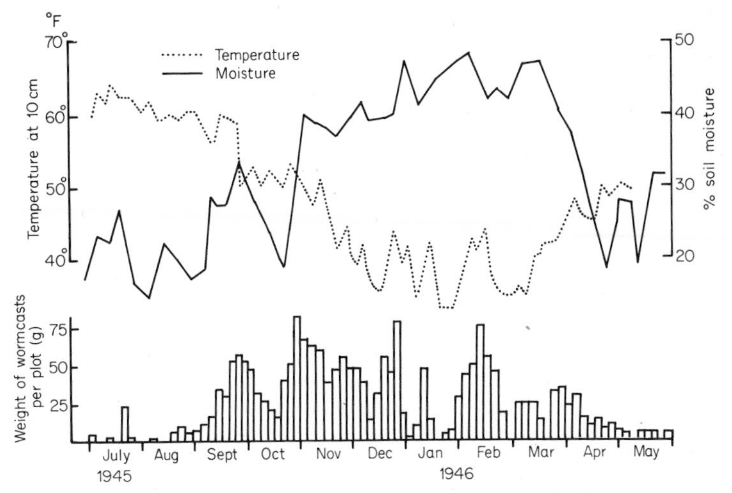

Fig. 39 The seasonal changes in wormcast production. (*After Evans and Guild, 1947*)

There are very many different forms of casts which are often typical of the species that produced them. They range from the small heterogeneous masses excreted by *A. longa*, and individual pellets produced by *Pheretima posthuma*, to short threads from *Perionyx millardi*. Some worms excrete a long, thick column of faeces which produces a hollow mound about 5 cm high and 2·5 cm in diameter. Ljungström and Reinecke (1969) described round hollows (1 m diameter, 30–100 cm deep) called 'kommetjies', in South Africa. They believed these to be formed by very large species of *Microchaetus* casting round their burrow margins (Fig. 40).

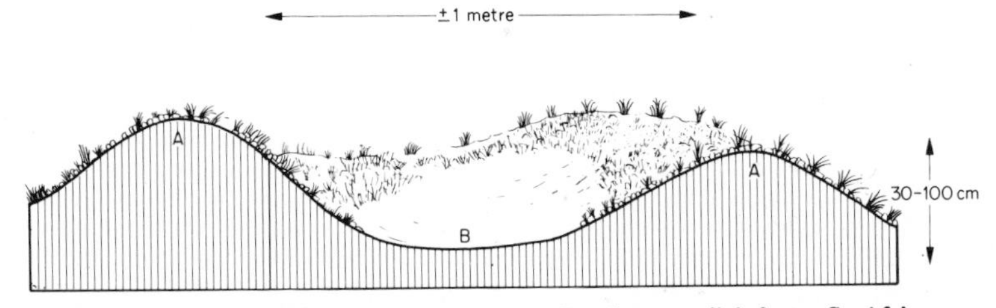

Fig. 40 Schematic vertical section diagram of a 'kommetjie' from S. Africa. A: walls with casting on the surface, B: bottom of a 'kommetjie' without casting. (*After Ljungström and Reinecke, 1969*)

Plate 7a Casts of *Hyperiodrilus africanus*

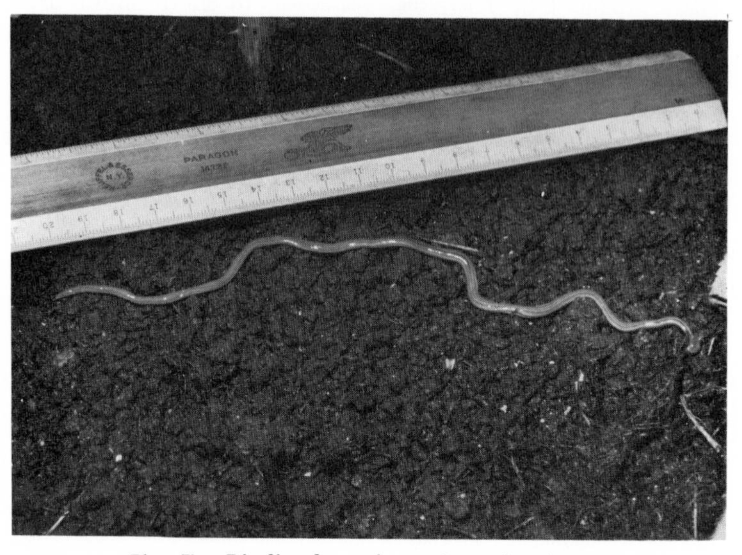

Plate 7b *Bipalium kewensis* a predator of earthworms

Earthworm casts can be very large; those of European worms seldom exceed 100 g, but African worms such as *Dichogaster jaculatrix* have casts in the form of red clay chimneys about 10–12 cm high and 4 cm in diameter (Baylis, 1915). The giant *Notoscolex* earthworms in Burma produce large tower-shaped casts 20–25 cm high and 4 cm diameter; one such cast weighed 1·6 kg. (Gates, 1961). *H. africanus* also produces large tower-like casts from 2·5–8·0 cm high and 1·0–2·0 cm in diameter (Madge, 1969) (Plate 7a). Both Madge and Gates (1961) agreed that casting by tropical species of earthworms is limited to the wet season. Most species of earthworms cast during the night, but *P. hupeiensis* usually deposits its casts during daylight hours (Schread, 1952).

Earthworm casts contain more micro-organisms, inorganic minerals and organic matter in a form available to plants, than soil. Casts also contain enzymes such as proteases, amylases, lipase, cellulase and chitinase, which continue to disintegrate organic matter even after they have been excreted (Ghabbour, 1966).

Darwin (1881) estimated that the annual production of worm casts in English pastures was 18·7–40·3 tonnes per ha (7·5–16·1 tons per acre), which is equivalent to a soil layer 5 mm deep being deposited annually. Guild (1955) calculated that 27 tonnes per ha were produced in another English pasture. Similar estimates for the amounts of casts deposited per annum in other parts of Europe

TABLE 7

Production of worm casts on Highfield bare fallow and adjoining permanent grass

	Bare fallow		Permanent grass	
Date	No. of casts	Wt. of casts (g)	No. of casts	Wt. of casts (g)
6–12 April	1·4	2·1	20·5	56·7
12–26 April	5·0	10·3	17·4	37·7
26 April–11 May	5·7	11·5	15·4	55·0
TOTAL	12·1	24·3	53·3	149·4

(*Lofty, unpublished data*)

range from 5·0–7·5 tonnes per ha for Germany (Kollmannsperger, 1934) to 75–100 tonnes per ha for Zurich (Stöckli, 1928).

In the tropics, even larger amounts have been reported, ranging from 50 tonnes per ha in Ghana (Nye, 1955), to as much as 2,100 tonnes per ha in the Cameroons (Kollmannsperger, 1956), and 2,600 tonnes per ha in the Nile Valley (Beauge, 1912).

The amounts of soil actually turned over by worms may be even greater, because some species void their casts underground (Evans, 1948). Nevertheless, the presence of casts is a good indication of the amount of earthworm activity; usually many more casts are found on the surface of pasture than on arable soils (Table 7).

5.7 Species associations

Certain earthworm species tend to be associated with one another. Usually, such associations result from some characteristic of the habitat, for instance, *L. terrestris*, *A. longa*, *A. caliginosa* and *O. cyaneum* are characteristic pasture species in England, although they are not the only species that occur in pasture. Similarly, there are rarely more than four species in peaty soils, and these are usually small worms. *Calluna* heath and coniferous mor soils often contain only two species, *D. octaedra* and *B. eiseni*. In woodland mull soils *A. rosea*, *A. longa*, *A. caliginosa*, *A. chlorotica*, *O. cyaneum*, *L. castaneus*, *L. terrestris* and *L. rubellus* are commonly found together. *D. rubida* was consistently associated with *Bimastos zeteki* in Michigan woodlands (Murchie, 1960). There are many such casual associations of species which are too numerous to list.

In Scotland no more than seven to ten species occurred in any one habitat, and *A. caliginosa* and *A. longa* usually seemed to be the dominant species, as many as 50% of the total number of worms belonging to these species. There seemed to be no relation between the age of a pasture and the number of species it contained (Guild, 1951). Nevertheless, the previous agricultural history of a grass or arable field is an important factor in determining which species are present. For instance, an old permanent pasture had many *A. nocturna* and rather fewer *A. caliginosa*, but ploughing and reseeding to grass, after one or two years in arable cultivation, favoured increases in numbers of *A. rosea* (Evans and Guild, 1948). *A. chlorotica* is often the dominant species in arable fields, and remains so for several years after arable land is reseeded with grass, eventually being replaced by more common pasture species.

Some associations of species are almost a form of commensalism and independent of the habitat. For instance, Lukose (1960) reported an association between the giant earthworm, *D. grandis* (Moniligastridae) and worms of a megascolecid species (11–32 mm long), which were found crawling over individuals of the larger species. When separated, the smaller worms always returned to the host worm. Baylis (1914) reported a similar association between an enchytraeid worm *Aspidodrilus* and a large species of earthworm, and Cernositov (1928) between an enchytraeid worm *Fridericia parasitica* and the earthworm *Allolobophora robusta.* A small lumbricid, *Dendrobaena mammalis*, sometimes lives commensally in the burrows of *A. terrestris* and *A. longa* (Saussey, 1957). A species of megascolecid worm, *Notoscolex termiticola*, in Ceylon, lives in termite mounds. The worm secretes a milky fluid which may attract or repel the termite (Escherich, 1911).

5.8 Predators and parasites

There is not a great deal of information about the relationships between earthworms and their more important predators and parasites. They are preyed upon by very many species of birds, by badgers and shrews, and especially by moles. Moles tunnel in search of earthworms, catch them and store them in caches until required as food; they ensure that the worms do not escape, by biting three to five segments from their anterior ends (Evans, 1948). It is probable that many other vertebrates also feed on earthworms, for instance, Ljungström and Reinecke (1969) reported that in South Africa, giant microchaetid worms are attacked by night adders (*Causus rhombeatus*).

A variety of predatory invertebrates also prey on earthworms. Many species of carabid and staphylinid beetles and their larvae attack earthworms; for instance, the staphylinid, *Quedius* (*Microsaurus*) *mesomelinus* fed on immature and mature individuals of *Eisenia foetida* under experimental conditions (McLeod, 1954). Centipedes also attack earthworms, and three species of carnivorous slugs, *Testacella scutulum*, *T. haliotidea* and *T. maugei* prey chiefly on earthworms, each individual eating about one worm per week. A few species of leeches, including the British species *Trocheta subviridis*, also feed on earthworms. A species of flatworm, *Bipalium kewensis*, attacks earthworms in greenhouse soils and is probably an important predator in warmer climates (Plate 7b).

Earthworms have many internal parasites including Protozoa, Platyhelminthes, Rotatoria, nematodes, and dipterous larvae. Bacteria that have been reported in earthworms include *Spirochaeta* and *Bacillus botulinus*, but little is known of their effects. No viruses have yet been recorded from earthworms. The most common and probably the most important protozoan parasites are the Gregarina. These have been found in many different parts of the bodies of earthworms, including the alimentary canal, coelom, blood system, testes, spermathecae, seminal vesicles and even in the cocoons. The records of species found in earthworms are too extensive to be included in full, but the genera of gregarines that have been found in earthworms include: *Distichopus*, *Monocystis*, *Rhyncocystis*, *Nematocystis*, *Echinocystis*, *Aikinetocystis*, *Grayallia*, *Nellocystis*, *Craterocystis* and *Pleurocystis* (Stephenson, 1930; Stolte, 1962).

A number of ciliate protozoa also infest the bodies of earthworms, although few cause any serious harm to the worms. The genera recorded include *Anoplophrya*, *Maupasella*, *Parabursaria*, *Hoplitophrya*, *Plagiotoma* and *Metaradiophrya* (Stolte, 1962). Other protozoa that have been found in the bodies of terrestrial earthworms include: *Myxocystis*, *Sphaeractinomyzon* and *Thelohania*.

There are several instance of platyhelminth worms being found in the bodies of earthworms. The cysticercoid stage of *Taenia cuneata* has been found in *E. foetida* and in the wall of the alimentary canal of a *Pheretima* sp. (Stephenson, 1930). The larvae of *Polycercus* have been found in the tissues of *L. terrestris* (Haswell and Hill, 1894).

Many nematodes occur in the tissues of earthworms; few seem to cause serious damage, and often the worm is merely acting as an intermediate host for them. (See Chapter 8, section 4.) Nematode genera that have been found in worms include: *Rhabditis*, *Heterakis*, *Syngamus*, *Dicelis*, *Stephanurus*, *Metastrongylus* (Stolte, 1962), *Spiroptera*, *Synoecnema* and *Diporochaeta* (Stephenson, 1930).

The larvae of muscoid flies parasitize earthworms, for instance, the cluster fly, *Pollenia rudis*, is a major parasite of earthworms in the United States. A worm may harbour one to four larvae but only one larva becomes fully grown and as it grows it progressively destroys the worm and eventually kills it. This fly was introduced into the United States from Europe many years ago, and its numbers have so increased that the adult flies are a considerable nuisance to human beings in many areas (Cockerell, 1924). Other flies that parasitize earthworms are *Onesia subalpina* (Takano and Nakamura, 1968), *O. sepulchralis*, *Sarcophaga haemorrhoidalis* (Keilin, 1915), *S. striata* and *S. carnaria* (Eberhandt, 1954).

5.9 Environmental factors

5.9.1 *pH*

It has been demonstrated that earthworms are very sensitive to the hydrogen-ion concentration (pH of aqueous solutions (see Chapter 4)), so it is not surprising that the pH of soil is sometimes a factor that limits the distribution, numbers and species of earthworms that live in any particular soil.

Several workers have stated that most species of earthworms prefer soils with a pH of about 7·0 (Arrhenius, 1921; Moore, 1922; Phillips, 1923; Salisbury, 1925; Allee *et al.*, 1930; Bodenheimer, 1935; Petrov, 1946). However, *L. terrestris* occurs in soils with a pH of 5·4 in Ohio, U.S.A. (Olson, 1928), and *A. caliginosa* in soils with a pH of 5·4–5·2 in Denmark, although there were few earthworms in soils with a pH below 4·3, except for one species, *D. octaedra*, which seemed to be acid-tolerant (Bornebusch, 1930). *Eisenia foetida* has been reported to prefer soils with a pH between 7·0 and 8·0 but, in contrast to this, certain tropical species of *Megascolex* thrive in acid soils from pH 4·5–4·7 (Bachelier, 1963), and *Bimastos lönnbergi* was numerous in soils between pH 4·7 and 5·1 (Wherry, 1924).

Satchell (1955) considered that *B. eiseni*, *D. octaedra* and *D. rubida* were acid-tolerant species, and *A. caliginosa*, *A. nocturna*, *A. chlorotica*, *A. longa* and *A. rosea* were acid-intolerant (Fig. 41). He considered that *L. terrestris* was not very sensitive to pH, and Guild (1951) agreed with this conclusion, although Richardson (1938) disagreed. Guild (1951) also confirmed that the relative abundance of *A. longa* and *A. caliginosa* was less in acid soils. Madge (1969) reported that the optimum pH for *H. africanus* lay between 5·6 and 9·2, so this species is acid-tolerant. *B. lönnbergi* and *Bimastos bed-*

dardi also seem to prefer acid soils (Wherry, 1924). An increase of pH from 7·25 to 8·25 was associated with a decrease in numbers of earthworms in the fourteen Egyptian soils studied by Duweini and Ghabbour (1965), so soils can also be too alkaline to favour earthworms. Jeanson-Luusinang (1961) observed in laboratory experiments that individuals of ***Eophila*** (***Allolobophora***) ***icterica*** that

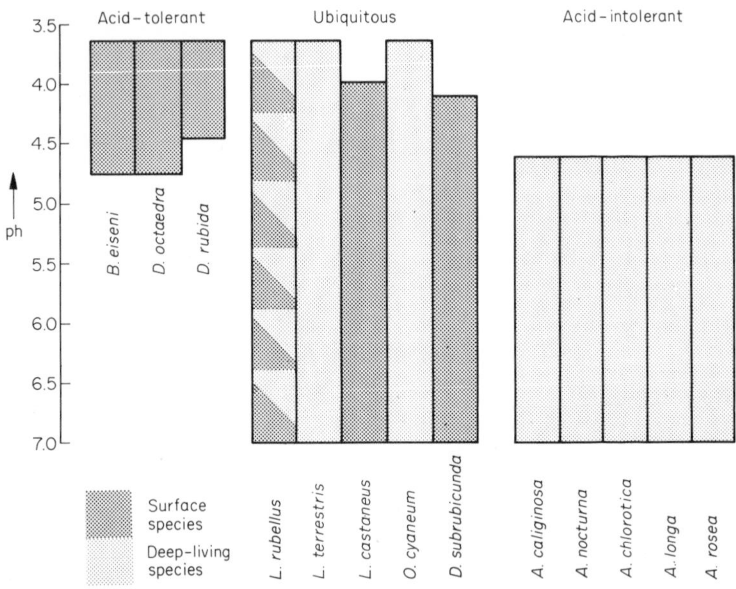

Fig. 41 Classification of earthworms as a function of the pH of litter. (*After Satchell, 1955*)

occurred in a field soil with a pH of 7·0, could tolerate soils with pH from 4·2 to 8·0, but were much more active at pH 8·0 than at pH 4·2; she speculated whether this might limit their penetration into deeper soil, which is normally more acid than surface soil.

Soil pH may also influence the numbers of worms that go into diapause. Doeksen and van Wingerden (1964) reported that when they put individuals of *A. caliginosa* into soils with pH's of 4·9, 5·4, 6·4, 7·6 and 8·6, the more acid the soil, the sooner worms went into diapause, and that they remained in diapause longest, in soil with a pH of 6·4.

Satchell (1955) took soil samples from plots with different fertilizer treatments, on a pasture experiment on Park Grass at Rothamsted. The pH values of the soils were 4·0, 4·1, 4·4, 5·0, 5·1, 5·6, 5·8, 6·9 and 7·0. He placed mature individuals of *A. chlorotica* on the surface of these soil samples, and studied their reactions and the time they took to bury themselves. In the three most acid soils, worms at first showed a violent avoiding reaction, twisting, jerking convulsively, and exuding coelomic fluid from their dorsal pores. They then extended to their full length and crawled about the soil surface, intermittently raising and waving the anterior segments. Activity gradually became sporadic and after one to two hours they lay motionless and became flaccid. After twenty-one hours, fifty-eight out of sixty worms exposed to pH below 4°4 were dead.

It has been reported by several workers that earthworm casts are usually more neutral than the soil in which the worms live [Salisbury, 1925; Puh, 1941; Stöckli, 1949; Dotterweich, 1933; Finck, 1952; Nye, 1955]. It has been suggested that one explanation for this is that earthworms neutralize soil as it passes through them, by secretions of the calciferous glands. This is now doubted, and a more probable explanation is that the soil is neutralized by secretions from the intestine, and by the ammonia which is excreted. How important this change of pH is to agricultural soils is unknown.

5.9.2 *Moisture*

Water constitutes 75–90% of the body weight of earthworms (Grant, 1955), so prevention of water loss is a major problem of earthworm survival. Nevertheless, they have considerable ability to survive adverse moisture conditions, either by moving to a more suitable area or by aestivating; if they cannot avoid dry soil they can survive the loss of a large part of the total water content of their bodies; *L. terrestris* can lose 70% and *A. chlorotica* 75% of their total body water and still survive (Roots, 1956).

Prolonged droughts markedly decrease numbers of earthworms and it may take as long as two years for populations to recover when conditions become favourable again. One reason for this seems to be that the fecundity of earthworms is greatly influenced by moisture (Table 8). Gerard (1960) showed that some species can withstand dry conditions better than others; for instance, *L. terrestris* survived as well in non-irrigated plots as in irrigated ones, whereas *A. chlorotica*, *A. caliginosa* and *A. rosea* did not survive in the non-irrigated plots. Olson (1928) surveyed areas of Ohio, U.S.A., for earthworms, and reported that the largest numbers of earthworms occurred in soils containing between 12% and 30% moisture. Duweini and Ghabbour (1965), who investigated the survival of *A. caliginosa* in relation to soil moisture, reported that in soils with 5–85% gravel and sand, an increase in moisture content of from 15% to 34% was usually associated with an increase in numbers of *A. caliginosa*, but above 34% extra moisture had no effect.

TABLE 8

The effect of soil moisture on cocoon production by A. chlorotica

Site	Bones Close						
Moisture content of soil	11	13·5	21	28	35·5	42·5	S.E. of difference
Mean No. of cocoons produced by five worms	0	0	8·6	13·6	8·8	6·6	0·94
Site	**Westfield**						
Moisture content of soil		1·6	24·5	33	42	50	S.E. of difference
Mean no. of cocoons produced by five worms		0	0·6	8·4	9·4	3·0	0·93

(*From Evans and Guild, 1948*)

Earthworms that migrate to deeper soil when the surface soil is too dry include *E. foetida* and *P. hupeiensis*, whereas *A. caliginosa*, which does not, is a species that can withstand desiccation better. Alternatively, lack of moisture can cause earthworms to become quiescent or go into diapause; for instance, when individuals of *A. caliginosa* were kept in soil that was dried slowly in the laboratory, they went into diapause, but when kept in moist soil, they remained active for eighteen months (Gerard, 1960).

When individuals of *A. caliginosa* and *A. longa* were offered a choice between water-saturated and air-filled soil, most worms chose the latter, but a few chose the saturated soil. *L. terrestris* and *D. subrubicunda* never occur in flooded soils although they have survived under water for several months in laboratory experiments.

Madge (1969) placed earthworms (*H. africanus*) in moisture gradients, and reported that they preferred soil with between 12·5% and 17·2% moisture. Soil with a moisture content of about 23·3% appeared to be optimum for them to produce casts. Earthworms are much more active in moist soils than dry ones, and during periods of much rain, individuals of some species such as *L. terrestris* come out on to the soil surface at night.

Many species of earthworms can survive long periods submerged in water, for instance, *A. chlorotica, A. longa, D. subrubicunda, L. rubellus* and *L. terrestris* were all able to survive from thirty-one to fifty weeks in soil totally submerged below aerated water. These species, and *A. caliginosa*, were also able to survive for ten to twenty weeks in aerated water without food. *H. africanus* could survive for more than nine weeks submerged (Madge, 1969). The main problem during submergence is the uptake of water into the worm, but most species have adequate means of overcoming this (see Chapter 4). Cocoons can also hatch under water, and the young worms feed and grow although totally immersed (Roots, 1956).

5.9.3 *Temperature*

The activity, metabolism, growth, respiration and reproduction of earthworms are all greatly influenced by temperature. Fecundity is affected very much by different temperatures; for instance, the numbers of cocoons produced by *A. caliginosa* and certain other lumbricid species quadrupled over the range 6–16°C (Evans and Guild, 1948) (Fig. 42). Cocoons also hatch sooner at higher temperatures, for example, cocoons of *A. chlorotica* hatched in 36 days at 20°C, 49 days at 15°C and 112 days at 10°C when there was adequate moisture (Gerard, 1960). Thus, during very cold weather few cocoons are likely to hatch, which is a useful feature for survival of the species, because newly hatched worms would be unlikely to survive very cold spells.

The growth period from hatching to sexual maturity is also dependent on temperature, for instance, *A. chlorotica* took 29–42 weeks to mature in an unheated cellar (Evans and Guild, 1948), 17–19 weeks at 15°C (Graff, 1953) and 13 weeks at 18°C (Michon, 1954). *E. foetida* took $9\frac{1}{2}$ weeks to mature at 18°C and only $6\frac{1}{2}$ weeks at 28°C (Michon, 1954).

Temperature also greatly affects the activity of earthworms and hence their metabolism and respiration. Kollmannsperger (1955) found that the number of worms on the soil surface at night was positively correlated with temperature, and reported that the

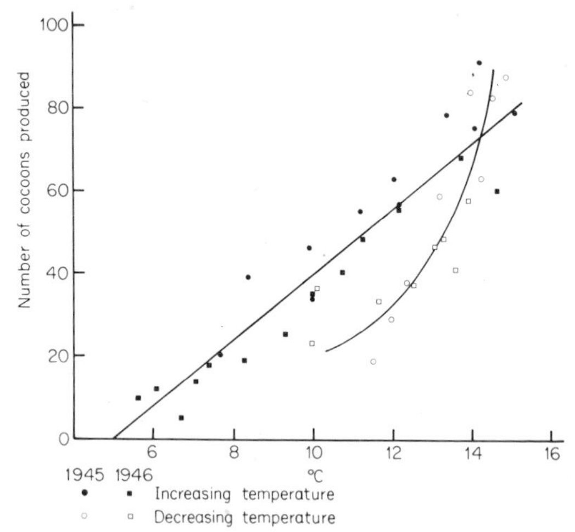

Fig. 42 Relationship between cocoon production by *L. rubellus* and increasing and decreasing temperature.
(*After Evans and Guild, 1948*)

optimum temperature for activity was 10·5°C. Satchell (1967) concluded that the most suitable conditions for activity of earthworms on the surface were nights when soil temperatures did not exceed 10·5°C, grass–air temperatures were above 2°C, and there had been some rain during the previous four days.

Temperature also affects the numbers of leaves buried by *L. terrestris* (Table 9) and this can be used as an index of earthworm activity.

The temperature at which earthworms thrive best and which they prefer is not necessarily that at which they grow fastest or are most active. Most of the evidence suggests a temperature preferendum for

L. terrestris of about 10°C, which is rather low compared with those for other invertebrates.

Earthworms can be killed by extreme temperatures. For instance, it has been suggested that earthworm populations in arable soils in the United States may be destroyed by frost (Hopp, 1947), but in pasture or woodlands it is unlikely that the soil would freeze deep enough to affect populations of most species. The upper lethal

TABLE 9

Effect of temperature on leaf burial
No. of leaves buried by 20 worms in 4 successive months

	0°C	5°C	10°C	15°C
December	0	25	35	67
January	0	33	31	30
February	0	53	63	33
March	0	67	75	44
TOTAL	0	178	204	174

(*Lofty, unpublished data*)

temperature for earthworms is lower than for many other invertebrates, being 28°C for *L. terrestris* (Wolf, 1938), 26°C for *A. caliginosa* and 25°C for *E. foetida* (Grant, 1955), although other workers have given 33·3°C as the lethal temperature for this last species, and 25·7°C for *A. longa.* Although the temperatures of the surface soil are occasionally higher than these, earthworms may still survive because they can maintain their body temperatures lower than that of the surroundings by evaporating water from the surface of their bodies (Hogben and Kirk, 1944).

Earthworms can migrate away from soil at unsuitable temperatures; in experiments in which individuals of *A. caliginosa* could choose between soils at different temperatures, they preferred soil from 10°C to 23°C, and individuals of *E. foetida*, soils from 16°C to 23°C (Grant, 1955). Dowdy (1944) reported that *Diplocardia* sp. moved down to lower levels at temperatures below 6°C. Madge (1969), who tested species of *H. africanus* in a temperature gradient, found that they aggregated between 23·9°C and 31·5°C,

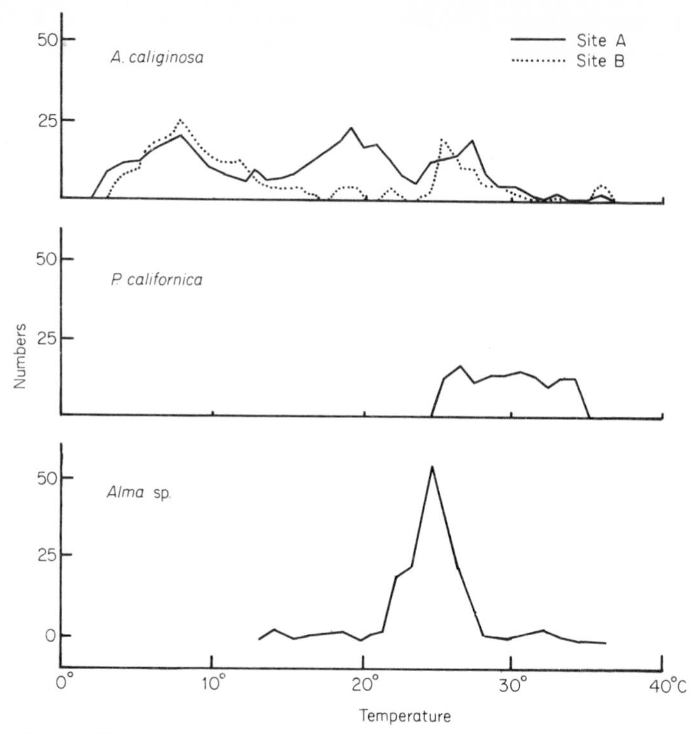

Fig. 43 Temperature preferences of *A. caliginosa P. californica* and *Alma sp.* (*After El-Duweini and Ghabbour, 1965*)

and temperatures above 34°C were avoided. These temperatures are higher than those which most other worms can survive, but this is a tropical species, and tolerance of high temperatures would aid survival. The upper thermal death point of this species was between 34°C and 38·5°C and the lower thermal death point 7·5°C.

Duweini and Ghabbour (1965) determined the preferred and lethal temperatures for *Pheretima californica* and a species of *Alma* (Fig. 43). The preferred temperatures for these species were 26–35°C and 24–26°C respectively, and the upper lethal temperatures 37°C and 38°C respectively. Worms could survive the higher temperatures better in moist air. Graff (1953) listed the optimum temperature for development of lumbricids (Table 26).

Grant (1955) showed that the temperature preferendum for *P. hupeiensis* was 15–23°C, for *E. foetida* 15·7–23·2°C and for *A. cali-*

ginosa 10–23·2°C. The upper lethal temperature, at which 50% of the test animals could be expected to die in 48 hours, was 24·9°C, 24·7°C and 26·3°C respectively for these three species after acclimatization at 22°C. He studied the effect of conditioning temperatures on the upper lethal levels of temperature for *P. hupeiensis*. Worms kept at 4°C, 9°C and 15°C had thermal death points of 19·4°C, 20·9°C and 22·7°C, with an average gain in heat tolerance of 0·3°C for every 1°C rise in conditioning temperature. Worms kept at 15°C took twelve days to acclimatize to a new environmental temperature of 22°C. Such mechanisms allow adjustment to seasonal changes in temperature.

5.9.4 *Aeration and Carbon dioxide*

There is little definite evidence that the soil oxygen tension affects the distribution of earthworms in soil, although Satchell (1967) stated that the distribution of *B. eiseni* and *D. octaedra* appeared to be limited in some sites by the minimum oxygen tensions occurring at certain seasons, but this was confused by factors such as pH, soil moisture content, amount of raw humus, plant cover and soil microflora status. He showed that there was some correlation between numbers of *B. eiseni* and oxidation-reduction potentials.

Some species can survive for long periods in very low oxygen tensions. Not much is known of typical oxygen tensions that occur in soils, but Boynton and Compton (1944) reported that there were oxygen tensions, in an orchard soil, below 10% for eleven weeks in the year at a depth of 90 cm, and for as long as six months at depths below 150 cm.

There is little evidence of the distribution of earthworms being affected by CO_2 concentrations in soil, and they do not seem to migrate in response to CO_2 concentrations, e.g. *E. foetida* did not respond to soil concentrations of CO_2 up to 25% (Shiraishi, 1954). The limits of CO_2 concentration in soil are normally between 0·01% and 11·5% (Russell, 1950), and earthworms can survive much greater concentrations than this, even up to 50% (see Chapter 4).

5.9.5 *Organic matter*

The distribution of organic matter in soil greatly influences the distribution of earthworms. Soils that are poor in organic matter do

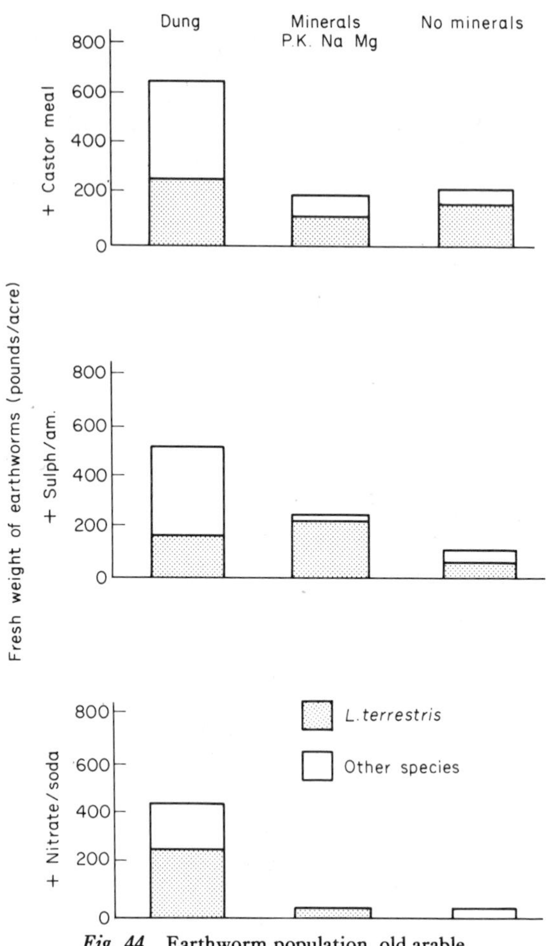

Fig. 44 Earthworm population, old arable.
(*Barnfield Mangolds*) *Kg/hectare* – (*after Lofty, unpublished*)

not usually support large numbers of earthworms. Conversely, if there are few earthworms, the decaying organic matter usually lies in a thick mat on the soil surface. Such mats of undisturbed organic matter occur in both woodlands (Richardson, 1938) and grassland (Raw, 1962). In irrigated pastures in New South Wales, Australia, that contained no earthworms, such mats of organic matter were up to 4 cm thick until earthworms were introduced experimentally.

Some species of earthworms are readily attracted to animal

droppings and dung. In experiments at Rothamsted which compared plots to which dung was added annually with plots which were left unmanured, large differences in earthworm populations were noted (Fig. 44). Plots that permanently grew wheat had earthworm populations that were three to four times greater in plots receiving 35 tonnes per ha of dung than they were in unmanured plots. The numbers in Park Grass, a permanent pasture fertilizer experiment, were three times greater in plots receiving 35 tonnes per ha of dung than in unmanured plots. In Barnfield, an arable field permanently growing mangolds, there were about fifteen times more earthworms in plots receiving dung annually than in unmanured plots (Table 10).

TABLE 10

Earthworm populations in plots with and without dung

Species	1. Grassland Park Grass, Rothamsted (Satchell, 1955)		2. Arable land Barnfield, Rothamsted (Lofty, unpublished)	
	Unmanured	Dung	Unmanured	Dung
L. terrestris	13·1	22·5	0·23	10·8
L. castaneus	16·0	59·6	—	—
A. caliginosa	2·9	8·0	0·8	15·4
A. chlorotica	1·6	—	3·2	44·6
A. rosea	10·0	21·3	—	0·23
A. longa	—	—	0·46	1·8
A. nocturna	1·3	18·9	—	—
O. cyaneum	6·9	24·5	—	—
TOTAL	51·8	154·8	4·69	72·83

Large amounts of dead roots and other organic matter in pasture usually coincide with large numbers of earthworms, and it is probably the gradual decrease in soil organic matter, when pasture is ploughed and used for arable crops, that leads to a decrease in earthworm populations.

Decaying leaves in woodlands are also a source of organic matter that usually favours earthworm multiplication. Earthworms can remove a large part of the annual leaf-fall in a woodland if populations are large and the litter is a leaf species that is palatable to the worms present. Increases in the organic carbon content of fourteen Egyptian soils were associated with increased numbers of earthworms (Duweini and Ghabbour, 1965).

5.9.6 *Soil type*

There have been very few studies of the direct influence of soil type, based on mechanical composition or earthworm populations. Guild (1948) made a survey of the main soil types in Scotland, and reported

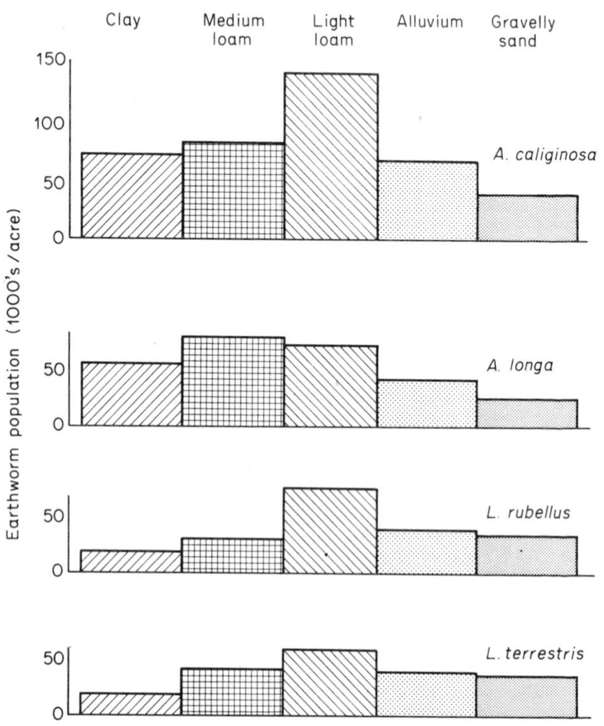

Fig. 45 Density of earthworm populations (thousands/acre) in various soil types in Scotland.
(*Adapted from Guild, 1948*)

that there were differences both in total numbers and relative numbers of each species (Fig. 45) (Table 11). Light and medium loams had higher total populations of worms than heavier clays or more open gravelly sands and alluvial soils. *A. caliginosa* was the dominant species in all soil types, but *A. longa* was less important in open soils, gravelly sands and alluvial soils. In a survey of the distribution of earthworm species in the Hebrides, Boyd (1957) compared the relative abundance of earthworm species in light soils with those in calcareous sand and dark peaty soils. Six species were more

TABLE 11

Relations of soil type to earthworm populations

Soil type	Population (Thousands/acre)	Population No/m²	No. of species
Light sandy	232·2	57	10
Gravelly loam	146·8	36	9
Light loam	256·8	63	8
Medium loam	226·1	56	9
Clay	163·8	40	9
Alluvium	179·8	44	9
Peaty acid soil	56·6	14	6
Shallow acid peat	24·6	6	5

(*From Guild, 1951*)

abundant on the light soils and six on the dark ones. In particular, *A. caliginosa* and *L. castaneus* were much more numerous in the light soils, and *B. eiseni* and *D. octaedra* in the dark soils.

Satchell (1967) reported that about eight characteristic species occur in mull sites in the Lake District, whereas in moder sites rarely more than four species occurred. In extreme mor soils, only two species, *D. octaedra* and *B. eiseni*, remained. Bachelier (1963) stated that worms were abundant in mull soils, but rare or absent in mor or moder soils.

A few small species of earthworms can survive in deserts and semi-deserts (Kubiena, 1953; Kollmannsperger, 1956), and some worms can inhabit the arid, cold soil of north-east Russia. It seems that although earthworms do better in good soils rather than poor ones, they can survive in many different kinds of soils providing there is adequate food and moisture.

5.9.7 *Food supply*

Earthworms can use a wide variety of organic materials for food, and even in adverse conditions extract sufficient nourishment from soil to survive. The kind and amount of food available influences not only the size of earthworm populations but also the species present, and their rate of growth and fecundity. Evans and Guild (1948) investigated the influence of food on cocoon production (Table 12), and showed clearly that more cocoons were produced by

TABLE 12

Mean number of cocoons produced by five earthworms in three months

Food	*A. chlorotica*	*L. castaneus*
Fodder	0·8	9·4
Oat straw	1·4	12·0
Bullock droppings	12·4	73·2
Sheep droppings	14·0	76·0

(*From Evans and Guild, 1948*)

worms that fed on decaying animal organic matter, than those fed on plant material. They also showed that earthworms fed on any nitrogen-rich diets grow faster and produce more cocoons than those with little nitrogen available (Evans and Guild, 1948). Barley (1959) fed individuals of *A. caliginosa* on different diets and reported that they grew at very different rates on these diets, growing fastest when fed on dung (Table 13). Worms consume large amounts of food, for instance, Guild (1955) calculated that worms of 0·1 g body weight eat as much as 80 mg of food per day per g of body weight of worm.

Guild (1955) stated that all species of earthworms prefer dung or succulent herbage to tree leaves, and that pine needles are preferred least of all; Barley (1959) corroborated this conclusion for *A. caliginosa*. Guild (1955) estimated that mature individuals of *A. longa* can ingest 35–40 g dry weight of dung per annum, *A. caliginosa* 20–24 g and *L. rubellus* 16–20 g.

Svendsen (1957) reported that whereas individuals of the pigmented species *Lumbricus festivus*, *L. rubellus*, *L. castaneus*, *D.*

octaedra, *D. rubida* and *B. eiseni* aggregated in dung, those of the unpigmented species *O. cyaneum*, *O. lacteum*, *A. caliginosa*, *A. chlorotica*, *A. longa*, *A. rosea* and *Eiseniella tetraedra* f. *typica* did not. Worms of the species *A. rosea* and *A. caliginosa* are not attracted to litter (Lingquist, 1941) but readily eat dung, and *A. caliginosa* will also feed on dead root tissues in pasture. Müller (1950) thought that *A. caliginosa* fed extensively on fungal mycelia and it seems that this species tends to be omnivorous. Individuals of *A. caliginosa* do not eat fallen leaves until these have become moist and brown (Barley, 1959), and they cannot maintain their body weight on dead roots; this tends to discount Waters' (1955) claim that dead roots are the main diet of this species.

TABLE 13

Changes in body weight of A. caliginosa when fed for forty days on various diets

Food	% change in wt
Angaston soil	−53
Phalaris roots	−26
Phalaris leaves	−26
Clover roots	−2
Clover leaves	+18
Dung, on surface	+71
Dung, incorporated	+111

(*From Barley, 1959*)

Most species of earthworms can distinguish between different kinds of forest litter. Darwin (1881) claimed that worms showed preference for leaves of particular shapes, but Satchell (1967) found that there was still an order of preference for certain leaf species, if uniform disks of a range of species of leaves were offered. Gast (1937) ascribed this preference to the mineral content of certain species of leaves but this has not been confirmed. Mangold (1951) believed that some species of leaves were unattractive to earthworms because of their bitter alkaloid or noxious aromatic content. Litter rich in protein is accepted more readily than that deficient in protein (Wittich, 1953), but protein content is often correlated with sugar content (Mangold, 1953; Laverack, 1960), which may be more important. Edwards and Heath (1963) buried leaf disks in nylon bags made from mesh with 7·0-mm apertures, so that earthworms could feed on them, and found that oak was preferred to beech and in another experiment (Heath *et al.*, 1966) stated that the order of preference for a range of different species of leaves buried in similar mesh bags was lettuce, kale, beet, elm, maize, lime, birch, oak and beech.

Leaves of larch, spruce, oak and beech, which are all comparatively unpalatable to earthworms, contain condensed tannins that are not found in dog's mercury, nettle, elderberry, ash and wych elm, all of which are more palatable (Brown *et al.*, 1963). King and Heath (1967) found that the amount of water-soluble polyphenols in litter was inversely proportional to the rate at which it was consumed, and that litter became much more palatable after a few weeks of weathering. In an excellent study of the palatability of litter to individuals of *L. terrestris*, Satchell (1967) showed that there was an inverse correlation between the palatability of litter and its total polyhydric phenol content, and a positive correlation with the amount of soluble carbohydrates. Such observations help to explain why the leaves of certain species of trees disappear from the soil surface faster than others and become broken down and incorporated into soil more readily.

There have been several different estimates of the amounts of food taken by earthworms and passed through their guts, but there is good agreement between different workers. Satchell (1967) calculated that individuals of *L. terrestris* pass 100–120 mg per g per day through their guts, and Crossley *et al.* (1971) calculated that a species of *Octolasium* had a soil intake of 1·2% of its live body weight per hour, which equals about 29% of its body weight per day or 290 mg per g per day. Barley (1959) calculated that *A. caliginosa* consumed 200–300 mg per g per day.

6. The role of earthworms in organic matter cycles

6.1 Fragmentation and breakdown

Plant organic material that reaches the soil is subject to many decomposing agents, including both micro-organisms and animals. Very soft plant and animal residues may be decomposed by the microflora but much organic matter, particularly the tougher plant leaves, stems and root material, does not breakdown without first being disintegrated by soil animals and acted upon by enzymes in their intestines. Earthworms have an important role in this initial process of the cycling of organic matter, because a few common species such as *L. terrestris* seem to be responsible for a large proportion of the fragmentation of litter in woodlands of the temperate zone. Soils with only few earthworms often have a well-developed layer of undecomposed organic matter lying on the soil surface. Edwards and Heath (1963) showed that in two sites earthworms consumed more oak and beech litter than all the other soil invertebrates together. In apple orchards *L. terrestris* removed more than 90% of the autumn leaf fall, during the course of the winter; this was calculated to be about 1·2 tonnes of dry weight of leaves per ha of orchard (Raw, 1962) (Fig. 46). The effectiveness of *L. terrestris* in initiating the fragmentation and incorporation of fallen apple leaves was vividly illustrated by comparing the soil profile and structure of an orchard with a large *L. terrestris* population, with one in which earthworms were almost totally absent (due to frequent and heavy spraying with a copper-based fungicide). The orchard with few earthworms had an accumulated surface mat, 1–4 cm thick, made up of leaf material in various stages of a very slow decomposition and sharply demarcated from the underlying soil, which had a poor crumb structure (Plate 8).

Many sorts of leaf litter are not acceptable to earthworms when they first fall on the ground, but require a period of weathering before they become palatable. It is believed that this weathering leaches water-soluble polyphenols from the leaves (Edwards and Heath, 1963; Heath and King, 1964; Satchell, 1967).

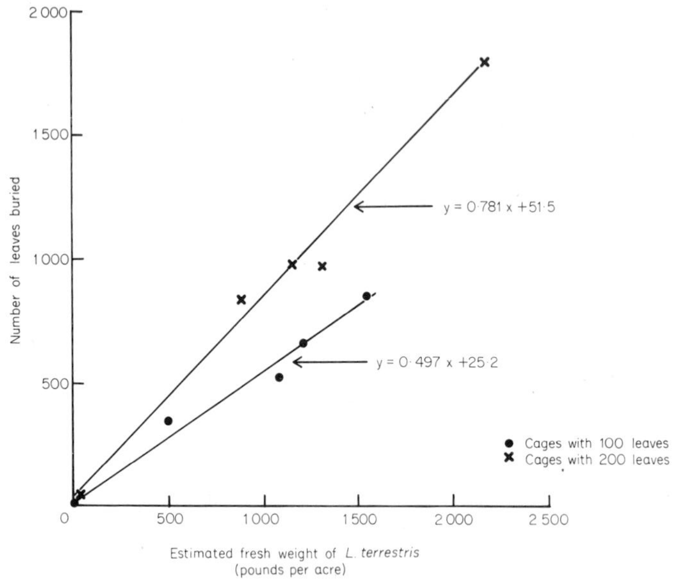

Fig. 46 Amount of leaf litter buried in relation to weight of *L. terrestris*. (*Adapted from Raw, 1962*)

When invertebrates were excluded from oak-leaf litter with naphthalene, after 140 days, one-half of the litter was decomposed when invertebrates were present, but only one-tenth this amount when they were excluded (Kurcheva, 1960). Edwards and Heath (1963) placed disks, cut from freshly fallen oak and beech leaves, in nylon bags of four different mesh sizes, which were then buried in woodland or old pasture soil. Only the bags with the largest mesh (7 mm) would allow the entry of earthworms as well as smaller animals. After one year, none of the fifty oak disks originally placed in each of the 7-mm-mesh bags remained intact, and 92% of the total oak-leaf material and 70% of the beech had been removed (Fig. 47). Earthworms ate not only the softer parts of the leaves but veins and ribs

Plate 8a Profile of orchard soil with no earthworms

Plate 8b Profile of orchard soil with burrowing earthworms

as well (Edwards and Heath, 1963). Other workers compared the disappearance of isolated samples of litter in 1-mm-mesh nylon bags with samples pinned under nylon net with the lower surface open to the soil. The leaves under the nylon net decomposed two to three times faster than those in bags, and this difference was mainly due to earthworms (Perel and Karpachevsky, 1966).

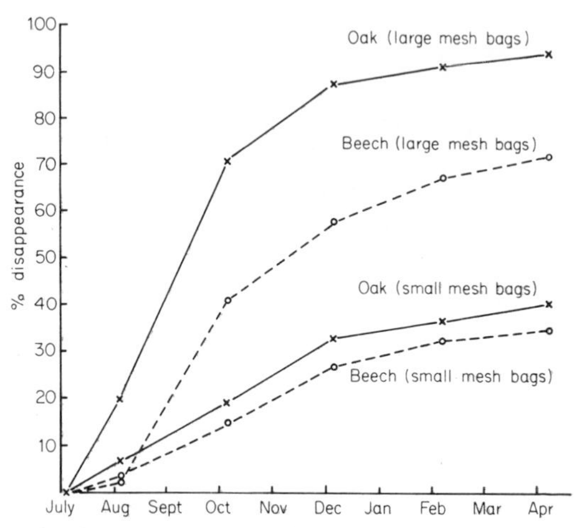

Fig. 47 Decomposition of leaf discs by soil animals. (*After Edwards and Heath, 1963*)

The rate of breakdown depends also upon the type of litter, so that beech leaves disappear much more slowly than oak leaves (Edwards and Heath, 1963), which in turn are more resistant to attack by earthworms than apple leaves (Raw, 1959). Elm, lime and birch disappear more rapidly than beech (Heath *et al*, 1966). Earthworms are much more attracted to moist litter material than to dry, and they are much more active in moist soil and litter. Earthworms can accelerate the decomposition of pine litter; when earthworms (*Dendrobaena octaedra*, *Dendrobaena attemsi*, *Dendrobaena rubida*, *Lumbricus rubellus*, *Bimastos eiseni* and *Allolobophora chlorotica*) were put in cultures containing pine litter, they fragmented and decomposed the pine needles (Heungens, 1969). The amount of fragmentation and decomposition was estimated from the settling of the litter material; after three months, the litter layer had settled by 13% when there were two earthworms per litre of litter, and by 26% when there were twelve earthworms per litre of litter. In New South Wales, pastures containing no earthworms also accumulated surface mats up to 4 cm thick, but these gradually disappeared after earthworms were introduced experimentally (Barley and Kleinig, 1964). Similar mats were found in plots regularly treated with ammonium sulphate on Park Grass, Rothamsted, and these plots had no earthworms.

6.2 Consumption, turnover, and humification

Earthworms seem able to consume very large amounts of litter, and the amount they turn over seems to be more dependent on the total amount of suitable organic matter available than on other factors. If physical soil conditions are suitable, the numbers of worms usually increase until food becomes a limiting factor.

Earthworms pass a mixture of organic and inorganic matter through their guts when feeding or burrowing, and in particular surface-feeding worms, such as *L. terrestris*, consume large amounts of organic matter. The smaller earthworms that feed on litter in woodlands, such as *L. castaneus* and *E. foetida*, produce casts that are almost entirely fragmented litter, whereas larger species such as *A. longa* and *A. caliginosa* consume a large proportion of soil, and there is less organic matter in their casts. Lumbricids, in old pasture land at Rothamsted, consumed between 50 and 90 tonnes of oven-dry soil per ha according to calculations by Evans (1948), but this was certainly an underestimate as the sampling method he used was inefficient. When individuals of *A. longa*, *A. caliginosa* and *L. rubellus* were fed on cow dung in cultures for two years, the average dry weight of dung each individual consumed during this time was 35–40 g, 20–24 g and 16–20 g respectively (Guild, 1955), and on this basis, the annual consumption of dung in the field by these species, at a population density of 120,000 adults per ha, would be 17–20 tonnes per ha, with a total estimated consumption for the whole population of about 25–30 tonnes per ha. Immature individuals of *A. caliginosa* consumed dung in culture, at a rate of 80 mg of oven-dried matter per g of fresh weight of worm (Barley, 1969), which was about twice the amount reported by Guild for this species. The amount of dung produced by dairy cattle (6–7·5 tonnes per ha) has been estimated as only one quarter of the amount that a typical earthworm population could consume (Satchell, 1967).

Crossley *et al.* (1970) calculated the rate of through-put of soil by *Octolasium* sp. in cultures of soil tagged with radiocaesium (^{137}Cs); they found that soil passed through the worms' guts at a rate of about 86 mg per day per worm, equivalent to 28·8% of the live weight of the earthworm. This compares well with calculations by Satchell (1967), who multiplied the weight of soil in dissected earthworms by an estimate of how rapidly food passes through their guts, that showed that individuals of *L. terrestris* consumed 100–120 mg or 10–30% of their live body weight per day and individuals of *A. longa*, 20% of their live body weight per day.

In an apple orchard, *L. terrestris* consumed the equivalent of 2,000 kg per ha of leaf litter between leaf fall and the end of February (98·6% of the total leaf fall) (Raw, 1962), and various workers have calculated the amount of different species of leaf litter eaten by earthworms. *L. rubellus* consumed 20·4 mg dry weight of hazel litter per worm (Franz and Leitenberger, 1948), six other species of worms consumed an average 27 mg of alder leaves per g fresh weight of worm (van Rhee, 1963), and *L. terrestris* consumed about 80 mg of elm leaves per g fresh weight of worm (Needham, 1957).

The weight of leaves that falls annually in woodlands has been estimated as varying from as little as 0·5 tonnes per ha per year in alpine and arctic forests, to 2·5–3·5 tonnes per ha per year in stable temperate forests, and as much as 5·5–15 tonnes per ha per year in tropical forests (Bray and Gorham, 1964), Satchell (1967) calculated that if a temperate deciduous woodland has a leaf fall of 3 tonnes per ha per year and if earthworms consume 27 mg per g of leaf litter per day, which is a reasonable average expectation, then they would consume the annual leaf fall in about three months. Madge (1966) calculated that in tropical forests in Nigeria, the litter fall was three to four times as much as in a temperate forest, and suggested that earthworms were the most important animals in fragmenting and incorporating it.

Even when organic material such as dung or litter is freely available to earthworms, many species also ingest large quantities of mineral soil. When individuals of *A. caliginosa* had unlimited quantities of litter available, they still ingested 200–300 mg of soil per gram of body weight per day, and the ingested mineral soil passed through the gut in about twenty hours (Barley, 1961).

The final process in organic matter decomposition is known as humification, and this is basically the breaking down of large particles of organic matter into a complex amorphous colloid containing phenolic materials. Only about one-quarter of the fresh organic matter becomes converted to humus. Much of the humification process is due to smaller soil organisms, such as microorganisms, mites, springtails and other arthropods, but is also accelerated by the passage of the organic material through the guts of earthworms feeding on decomposed organic matter together with mineral soil. Probably some of the final stages of humification are due to the intestinal microflora in the earthworms' gut, because most of the evidence indicates that the chemical processes of humification are caused more by the microflora than by the fauna. The full role of earthworms in the decomposition of the more resistant organic remains is still not clear and needs more study. The major contribution of earthworms seems to be in breaking up organic matter, combining it with soil particles and enhancing microbial activity when humification is well advanced. Nevertheless, earthworms are also important in mixing the humified material into the soil.

6.3 Nitrogen mineralization

Earthworms greatly increase soil fertility, and at least part of this must be due to the increased amounts of mineralized nitrogen that they make available for plant growth. There is no evidence that earthworms have any influence on the nitrogen-fixing bacteria (Day, 1950; Khambata and Bhatt, 1957), and it has been reported that the numbers of *Azotobacter* decrease when passing through earthworms (Ruschmann, 1953), although numbers may later increase in cast soil (Stöckli, 1928). There have been reports of increases of the amounts of nitrogen in soils in which earthworms were reared (Russell, 1910; Blancke and Giesecke, 1924; Lindquist, 1941) but this could be from decay of the bodies of dead worms. Earthworm corpses decay rapidly, and in one experiment they had disappeared completely from soil after two to three weeks at 12°C. Of the nitrogen added to the soil from the decomposed worm tissue, 25% was in the form of nitrate, 45% as ammonia, about 3% as soluble organic compounds, and the 27% that was unaccounted for probably consisted of undecomposed remains of setae and cuticle and microbial protein (Satchell, 1967).

The body of a worm contains up to 72% of its dry weight as protein (Lawrence and Millar, 1945), and it has been calculated that the body of a single dead worm can yield as much as 10 mg of nitrate. If these figures are accepted it can be calculated that a population of $3\frac{3}{4}$ million earthworms per ha could yield the equivalent of about 217 kg per ha of nitrate of soda. Obviously this is speculation, because only a small proportion of the total population dies at any one time. If, however, it is assumed that the average life of an earthworm is about one year, this would represent a possible annual addition of nitrate from the annual mortality of a typical earthworm population.

Earthworms consume large amounts of plant organic matter that contains considerable quantities of nitrogen, and much of this is returned to the soil in their excretions. It has been estimated that of the total nitrogen excreted by worms, about half is secreted as mucoproteins by gland cells in the epidermis, and half in the form of ammonia, urea, and possibly uric acid and allantoin, in a fluid urine excreted from the nephridiopores. The exact proportions of these constituents and the total amount of the nitrogen excreted depends upon the species of worm, and whether it is feeding or not. It seems likely from experimental studies that earthworms excrete less nitrogen when they are feeding than when they are not and are living only on their reserves (Needham, 1957; Cohen and Lewis, 1949). If this is accepted, it is likely that earthworms excrete more available nitrogen in dry weather, when they feed less, than in wet weather when they are active and feed more.

It has been suggested that very little nitrogen is excreted in the faeces of earthworms (Needham, 1957), but other workers have reported considerably more nitrogen in casts than in the surrounding soil (Lunt and Jacobson, 1944; Graff, 1971). When young worms, of the species *A. caliginosa*, were fed on soil containing finely ground plant litter, and their faeces and urine collected, about 6% of the non-available nitrogen ingested by the worms was excreted in forms available to plants (Barley and Jennings, 1959) (Table 14). The presence of worms in cultures of well-aerated moist soil increased the rate of oxygen consumed and the rate of accumulation of ammonium and nitrate during the early stages of decay.

Needham's figures for nitrogen excretion have been used by several workers, in calculations of total nitrogen turnover by earthworms, but it should be borne in mind that they assumed that little or no nitrogen was excreted with the faeces. For instance, from these figures, it was calculated that ley plots, that contained 126 g fresh

TABLE 14

Influence of A. caliginosa on decomposition of organic matter

	Oxygen consumed (30 days)		Nitrate and ammonium accumulated (50 days)	
	μl per g of medium	log value	ppm	log value
With worms	2,600	3·39	129	2·10
Without worms	2,190	3·32	105	2·01
Sig. diff. ($p = 0{\cdot}01$)	—	0·05	—	0·09

(*From Barley and Jennings, 1959*)

weight of earthworms per m², produced about 70 kg per ha of mineralizable nitrogen, although some of this may be consumed again by other earthworms (Heath, 1965). This was about half of the amount of mineralizable nitrogen that occurred in the soil of the plots before they were ploughed and put down to ley. There was a strong correlation between the numbers of earthworms in the plots and the mineralizable nitrogen in the soil at the time of ploughing. However, these calculations are based only on the amounts of nitrogen excreted and do not allow for that returned in dead bodies, so Satchell (1957) made calculations that took into account both nitrogen excreted and nitrogen returned in dead bodies. He determined the size of a *L. terrestris* population in an ash/oak wood and, from data on the rate of growth of this species, he calculated that the weight of earthworm tissue produced in one year was about 364 g per m². The nitrogen content of *L. terrestris* is about 1·75% and on this basis he calculated that the nitrogen returned to the soil in dead worms would be about 6–7 g per m^2 per annum. This is in addition to the nitrogen excreted by worms, and using Needham's figures for nitrogen excretion, he estimated that the joint yield of nitrogen from excreta and dead worms, would be of the order of 100 kg per ha per annum. This is about double the amount of nitrogen needed by agricultural crops, and several times the amount used by trees (Satchell, 1963). Lakhani and Satchell (1970) later estimated that tissue production was 50–76 g per m^2 per annum, so this may be an overestimate. Such figures suggest that earthworms in woodlands can consume most of the nitrogen being added with leaf litter and, when it is considered that they are not the only animals using this for food, it seems that the availability of food is probably the main factor limiting populations of earthworms.

6.4 Effects on the C : N ratio

The ratio of carbon to nitrogen (C : N ratio) in organic matter added to soil, is of importance because plants cannot assimilate mineral nitrogen unless this ratio is of the order of 20 : 1 or lower. The C : N ratio of freshly fallen litter is much higher than this, being 24·9 : 1 for elm, 27·6 : 1 for ash, 38·2 : 1 for lime, 42·0 : 1 for oak, 43·5 : 1 for birch, 54·0 : 1 for rowan and 90·6 : 1 for Scots pine (Wittich, 1953). Succulent leaf material often has small C : N ratios whereas tougher tree leaves with a high percentage of resistant constituents such as cellulose and lignin, that are unpalatable to earthworms and other litter animals, often have high C : N ratios (Witkamp, 1966).

During the process of litter breakdown and decomposition, the C : N ratio of the litter decreases progressively from ratios of the order of those given above for tree leaves; this means that none of the nitrogen they contain is directly available to living plants until the ratio falls to about 20 : 1, when nitrogen can be directly taken up by plants (Edwards *et al.*, 1970). Earthworms feeding on litter gradually lower its C : N ratio as they break the material down during their metabolism. This lowering of the C : N ratio is achieved mainly by combustion of carbon during respiration. To assess the role of earthworms in lowering the C : N ratio, the consumption of carbon must be measured, and this can be done approximately by measuring the respiration. There are disadvantages to this method, in that

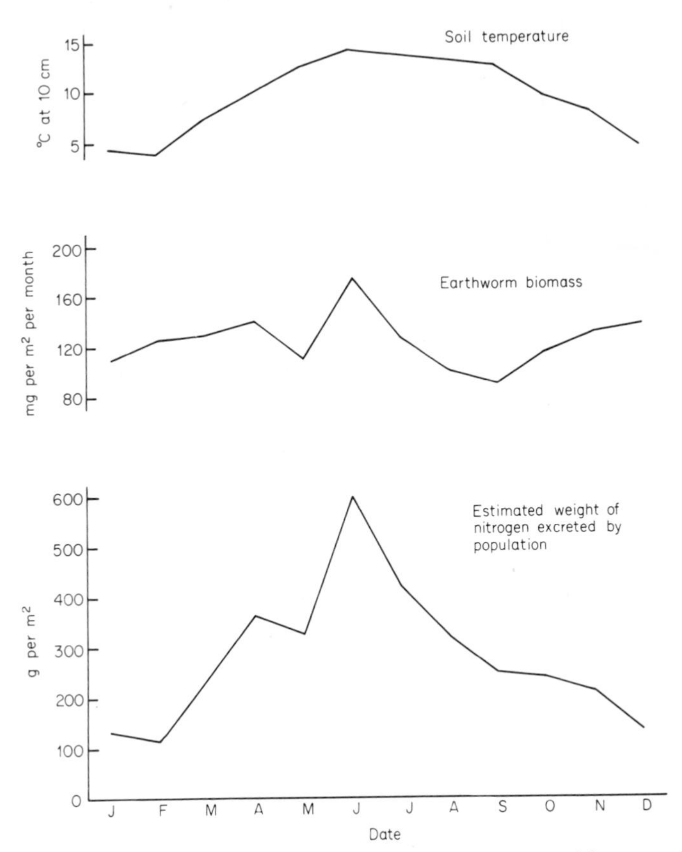

Fig. 48 Estimated weight of nitrogen excreted by a population of *L. terrestris* at Merlewood Lodge Wood in 1960.
(*After Satchell, 1963*)

laboratory tests do not always reflect respiration in the field, and because respiration is affected considerably by diurnal activity (Fig. 49), oxygen and carbon dioxide tensions, exposure to light, temperature, and because earthworms can accumulate oxygen debts. (See Chapter 4.)

It is not proposed to go into details of the calculations but it has been calculated for a population of the species *A. caliginosa* in Australia, that worms were responsible for only 4% of the total carbon consumption (Barley and Kleinig, 1964), and in two English woodlands, *L. terrestris* was responsible for only 8% of the total carbon consumption (Satchell, 1967). This was assuming that the consumption of 22·9 1 per m^2 of oxygen was equivalent to a carbon consumption of 118·6 kg per ha, and that 3,000 kg of litter that was 50% carbon fell per ha. These estimates of carbon consumption are probably too low because, when respiration was measured, worms were inactive. Nevertheless, these rates of carbon consumption by earthworms are too small to lower the C : N ratio sufficiently to

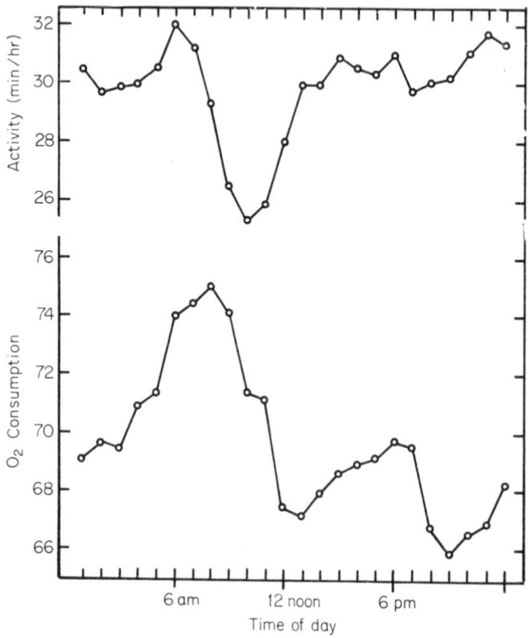

Fig. 49 Diurnal activity in *L. terrestris.* (*After Satchell, 1967*)

make nitrogen available to plants, and Satchell considered that a figure of 12% for the carbon respired by earthworms would be more realistic. Nitrogen production varies seasonally (Satchell, 1963) (Fig. 48).

6.5 Effect on available mineral nutrients

Most workers who have examined the available mineral nutrients in both earthworm casts, and in soil with many earthworms, have reported that they generally have a higher base-exchange capacity, and more exchangeable calcium, magnesium and potassium, and available phosphorous, than soils without earthworms. This was demonstrated by Lunt and Jacobson (1944) (Table 15) and Graff

TABLE 15

Comparison of the available mineral elements in the casts of earthworms and in the upper layers of a ploughed soil in Connecticut, U.S.A.

	Earthworm excreta	Depth of soil layer (0–15 cm)	Depth of soil layer (20–40 cm)
Loss by ignition (%)	13·1	9·8	4·9
Carbon/nitrogen ratio	14·7	13·8	13·8
Nitrate nitrogen (ppm)	21·9	4·7	1·7
Calcium:			
Total (%)	1·19	0·88	0·91
Exchangeable (ppm)	2,793	1,993	481
Exchangeable calcium/total calcium (%)	25·6	24·4	6·1
Magnesium:			
Total (%)	0·545	0·511	0·548
Exchangeable (ppm)	492	162	69
Exchangeable magnesium/total magnesium (%)	9·19	3·24	1·29
Phosphorus available (ppm)	150	20·8	8·3
Potassium available (ppm)	358	32·0	27·0
pH	7·00	6·36	6·05

(*From Lunt and Jacobson, 1944*)

(1971) (Table 16), and has been confirmed by other workers (Powers and Bollen, 1935; Puh, 1941; Stöckli, 1949; Ponomareva, 1950; Finck, 1952; Nye, 1955). Earthworms have also been shown to be important in New Zealand pastures that are rich in total molybdenum but poor in available molybdenum; the introduction of European species of earthworms greatly increased the amounts of molybdenum available for plants. Earthworms have greatly increased the amounts of soluble and available nitrogen, phosphorus and potassium, in a rather poor brown earth soil (Graff, 1972), and

TABLE 16

Mineral elements in soil and worm casts
(mg per 100 g dry wt)

	Casts	Soil
C	8,550	3,925
N	536	350
C : N ratio	16	11
P (aqua regia)	102	68
P (lactate-soluble)	13·7	2·2
K (soda extract)	1,097	799
K (lactate-soluble)	44·6	7·0
pH (H_2O)	5·8	5·0
pH (KC1)	5·4	4·0

(*From Graff, 1972*)

Nye (1955) reported that the casts of the tropical worm ***Hippopera nigeriae*** were much richer in exchangeable Ca and Mg than the top 15 cm of soil in forests in south-west Nigeria. Nijhanan and Kanwar (1952) studied the mineral content of earthworm casts in India, and stated that there was more Ca in all casts than in soil, but that there was more avilable K, Mn and exchangeable Ca, Mg, K and Na in large casts than in either soil or small casts, and also that the amounts of these minerals were less in small casts than in the surrounding soil. Thus, all the available evidence is that earthworms make more mineral nutrients available for plant growth; and this may be important in improving soil fertility.

7. Earthworms and micro-organisms

7.1 Effects of earthworms on number of micro-organisms

It is generally agreed that the species of micro-organisms in the alimentary canal of earthworms are usually the same as those in the soils in which the worms live. For instance, Bassalik (1913) isolated more than fifty species of bacteria from the alimentary canal of *Lumbricus terrestris*, but found none that differed from those in the soil from which the worms had been taken. This was confirmed for three species of earthworms by Parle (1963), who reported that most of the cellulose and chitin enzymes that occur in the alimentary canals of earthworms are secreted by the worms and not by symbiotic micro-organisms, as they are in some arthropods. From such observations, Satchell (1967) concluded that it was unlikely that earthworms have an indigenous gut microflora. Although the microflora of the alimentary canal is qualitatively similar to, or has fewer species than the surrounding soil, Stöckli (1928) found that there was a great increase in total numbers of bacteria and actinomycetes in the earthworm gut compared with those in the soil, numbers increasing exponentially from the anterior to the posterior portions of the gut (Table 17) (Parle, 1959 a).

Ponomareva (1953) stated that there is an increase in numbers of actinomycetes, pigmented bacteria and other bacteria of the *Bacillus cereus* group after passage through the earthworm intestine, and the same worker found that in soil spoiled by erosion, the number of bacteria in earthworm faeces was thirteen times as high as in the surrounding soil (Ponomareva, 1962). It has been suggested that the large numbers of bacteria ($5{\cdot}4 \times 10^6$ per g) that occur in the B layer of chernozem soil (depth 50–60 cm), can be explained by the burrowing activities of earthworms that penetrate this horizon and leave their casts. Teotia (1950) reported that worm casts had a bacterial count of 32·0 million per g compared with 6·0–9·0 million per g in the surrounding soil. There were fewer actinomycetes and fungi in the casts but more *Azotobacter* and other bacteria.

Zrazhevski (1957) observed that the density of the bacterial population, in turf soil without earthworms, was $2 \cdot 8 \times 10^6$ per g, whereas in the casts of worms added to this soil, it was $9 \cdot 8 \times 10^6$ per g, and in the same soil after a long colonization by worms, also

TABLE 17

Numbers of micro-organisms in different parts of the intestine of L. terrestris

	($\times 10^6$)		
	Fore gut	Mid gut	Hind gut
Actinomycetes	26	358	15,000
Bacteria	475	32,900	440,700

(*From Parle, 1959*)

$9 \cdot 8 \times 10^6$, or three times greater than initially. Kozlovskaya and Zhdannikova (1961) reported that not all species of earthworms have the same relationship with soil micro-organisms. They compared the microbial populations of the gut contents of *Octolasium lacteum* and *Lumbricus rubellus*. *O. lacteum* lives at a depth of 10–40 cm, and its gut contents had about the same density of bacteria as that in the soil in which it lives, although there were rather more spore-forming bacteria and actinomycetes, and fewer fluorescing bacteria in the gut contents than in the soil. By contrast, *L. rubellus* lives in the top 5 cm of soil, and the total density of bacteria in its gut is ten or more times larger than that in soil, depending very much on the food of the worm. The excreta of both species contain more fungi, actinomycetes, butyric acid-forming bacteria of the *Clostridium* type, and cellulose-decomposing bacteria. Decomposition of organic matter was much faster and more intensive in the casts than in the soil.

Atlavinyte and Lugauskas (1971) reported from pot tests, that earthworms increased the number of micro-organisms in soil as much as five times. Thus, there is good evidence that earthworms are important in inoculating the soil with micro-organisms and their casts are foci for dissemination of soil micro-organisms (Table 18) (Ghilarov, 1963). Many other workers have reported that the micro-floral population of cast soil is larger than that of the surrounding

TABLE 18

Numbers of micro-organisms in earthworm casts and soil

	(thousands per g)		
	Oak forest	Rye field	Grass field
Earthworm casts	740	3,430	3,940
Soil	450	2,530	2,000

(*From Ghilarov, 1963*)

soil (Bassalik, 1913, Zrazhevski, 1957; Went, 1963), especially in casts formed from soil containing added organic matter (Dawson, 1948).

However, there is evidence that some of the micro-organisms that are taken in with the soil are killed during their progress through the earthworm's alimentary canal. Aichberger (1914) reported that the crops, gizzards and intestines of earthworms contained few live organisms that did not possess a firm outer coat, and found no diatoms, desmids, blue-green algae, rhizopods or live yeasts. Dawson (1947) reported that the numbers of species of bacteria in soil that had passed through the gut of an earthworm, were decreased, whereas those of fungi seemed unaffected. Day (1950) stated that when soil heavily inoculated with *B. cereus* var. *mycoides* passed through the gut of *L. terrestris*, the numbers of bacilli decreased greatly, although a few survived in the gut, indicating that vegetative cells were destroyed, rather than spores. Two other bacteria, *Serratia marcessens* (Day, 1950) and *Escherichia coli* (Brüsewitz, 1959), that had been introduced into soil by inoculation, were killed after the soil had been ingested by *L. terrestris*. Khambata and Bhatt (1947) found that the bacillus *E. coli* was usually absent from the intestines of *Pheretima*, although these worms often live in soil that is regularly manured with human excreta, and they suggested that secretions in the intestine of the earthworms possibly prevented the growth of this and other pathogens. Kozlovskaya and Zhdannikova (1961) reported that the ratios of different groups of micro-organisms in soil differed from those in casts, so that the spore-forming bacteria and actinomycetes predominated in casts, and the numbers of *Bacillus idosus* and *B. cereus* were greater in casts than in soil, but those of *B. agglutinatus* were less.

Several workers have shown that at least some species of fungal spores can survive passage through the alimentary canal of earthworms. Hutchinson and Kamel (1956) isolated seventeen species of viable fungi from the alimentary canal and rectum of ten individuals of *L. terrestris*, and many of these fungal spores had relatively thin walls; they suggested that many more species of fungi would have survived if a larger number of earthworms had been sampled. Also, despite the limited number of species of fungi represented, some occurred consistently in worms from different areas, and at different times of the year. There were fewer microorganisms isolated from worms examined in midwinter, than in late autumn, which corresponds to the seasonal fluctuations in microbial populations in soil that have been reported by other workers. Day (1950) stated that the total number of bacteria, actinomycetes and fungi in casts was not consistently greater or less than in the soil in which the worms had been living. However, he used especially prepared soil in culture containers, and introduced the worms to compacted soil, so that the worms had to burrow by ingesting soil rather than by pushing through it. In these conditions, more soil passed through the gut than it would in normally feeding worms, and the short time soil was in the gut may have been insufficient to allow bacteria to multiply much. Other workers, e.g. Jeanson-Luuisinang (1963) and West, Brusewitz (1959), who reported little or no differences between microfloral populations of earthworm casts and soil, used soil to which was added a readily available organic energy source that would minimize any differences in the microflora of gut and casts. It seems clear that the microflora of earthworm casts is usually richer than that of soil.

The microbial populations of the earthworms' intestine begins to change when the gut contents are voided as casts, that are usually rich in ammonia and partially digested organic matter, and provide a good substrate for growth of micro-organisms. Such changes were first reported by Stöckli (1928), who estimated that the total microbial cell count in the earthworm casts doubled in the first week after they were deposited. Thereafter, for a further three weeks, numbers did not increase overall, although they fluctuated considerably during this period (Fig. 50). Parle (1963 b) reported that yeasts and fungi which occurred in the soil as spores, germinated as soon as

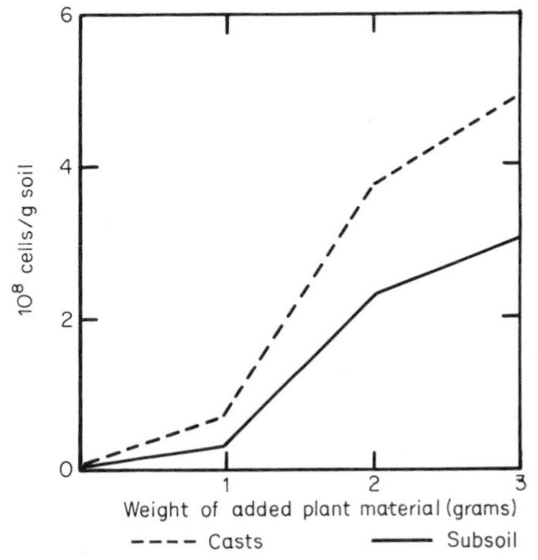

Fig. 50 Microbial cell content of faeces produced by *E. foetida* when kept in subsoil and fed on dried ground plant material.
(*After Brüsewitz, 1959*)

they were in worm casts and most hyphae were formed in casts fifteen days old. No consistent changes in numbers of actinomycetes or bacteria were found in old casts. Microbial activity, as shown by oxygen consumption, was stated to decline from the moment casts are produced (Parle, 1963), and the reason suggested for the inconsistency of a decline in oxygen consumption corresponding with an increase in microbial population, is that, as the casts age, an increasing proportion of the microflora forms resting stages.

The micro-organisms *Nocardia polychromogenes*, *Actinomyces* spp, and *Streptomyces coelicolor* which have been isolated from the gut contents and casts of earthworms, were reported by Ruschmann

(1953) to be particularly antagonistic towards the anaerobic spore-forming bacteria, and their antibiotic effect may result in an overall decline in respiratory exchange within earthworm casts. Other workers have noted that earthworms may produce antibiotic substances, for instance, it has been shown that growth of certain fungi, on soil in a petri dish, stopped whenever an earthworm was introduced (van der Bruel, 1964) and Ghabbour (1966) reported that when earthworms were placed in dilute glucose or glycine solutions, fungi did not grow until the earthworms died. Kobatake (1954) reported that earthworm extracts were anti-bacterial against several strains of non-acidfast pathogenic micro-organisms and saprophytic mycobacteria (a petroleum-ether extract was bacteriocidal at a dilution of 1 in 1,000 and bacteriostatic at 1 in 3,200).

Differences in the size of microfloral populations of cast soil and the surrounding field soil, may be due either to changes occurring in the earthworm's intestine, or because the selected food material ingested by the worm forms a richer substrate for microfloral activity, and it is not usually easy to determine which of these is the major factor involved. The size of the microfloral population in casts, certainly depends on the type and quantity of ingested plant material they contain, but as Parle (1959) showed, numbers of bacteria and actinomycetes contained in ingested material, increased up to 1,000-fold while passing through the earthworm gut, and oxygen consumption, which is an indicator of microbial activity, was still considerably higher in cast soil than in the surrounding soil, even fifty days after being excreted.

It has been suggested that microfloral activity in worm casts may have an important effect on soil crumb structure, by increasing the stability of worm-cast soil relative to that of non-cast soil. Many workers have shown that worm casts contain more water-stable aggregates than non-cast soil (see Chapter 6), but the way in which microflora contribute to this has not been explained, although one suggestion is that soil particles in worm casts are stabilized by accumulations of polysaccharide gums produced by the intestinal bacteria (Satchell, 1958).

Some species of protozoa seem to be essential to individuals of *Eisenia foetida*, for normal growth, because Miles (1963) showed that these worms are unable to reach sexual maturity without certain mobile protozoa in their food; these protozoa are normally abundant in the habitat of this worm, which lives in compost and manure heaps, and soils with a very high organic content. This may be one reason why this species disappears so rapidly when pasture soils are cultivated and a succession of arable crops grown.

7.2 Effects of earthworms on dispersal of micro-organisms

Earthworms may be instrumental in dispersing fungi or bacteria by ingesting their spores in one place and excreting them elsewhere. Many thick- and thin-walled spores lose little viability during passage through the intestines of earthworms (Hoffman and Purdy, 1964), e.g. spores of dwarf bunt (*Tilletia controversa*) lose no viability in the worms. It has also been suggested that earthworms disperse spores of harmful fungi such as *Pythium* (Baweja, 1939) and *Fusarium* (Khambata and Bhatt, 1957). When Hutchinson and Kamel (1956) inoculated sterilized soil with spores of several different species of fungi, the rate of spread of the fungi through the soil was much greater when there were earthworms, than if they were absent.

Earthworms may also have an adverse affect upon the spread of fungi, for example, the ascospores of *Ventura inaequalis* (Cooke) Wint, which causes apple scab, are released from perithecia on over-wintering dead leaves lying on the soil surface in the spring, and these infest the new growth (Hirst *et al.*, 1955). However, a large population of *L. terrestris* removes many of these leaves from the soil surface during the winter, thus preventing at least a proportion of the ascospores from being able to infect trees. It seems clear that earthworms must be very important vectors of plant pathogens but much more work is required to assess this importance.

7.3 Stimulation of microbial decomposition

The decomposition of organic material in the soil is accelerated when simple nitrogenous compounds are added to soil (Tenney and Waksman, 1929; Harmsen and van Schreven, 1955), particularly if the organic material is poor in nitrogen. Therefore, because the excreted cast material from earthworms is usually rich in nitrogenous compounds, large numbers of earthworms in an organic soil not only help to decompose organic material in the soil by ingestion, disintegration and transport, but their waste products may also stimulate other microbial decomposition processes as well. Barley and Jennings (1959) added grass and clover litter and dung pellets to soil culture pots with and without earthworms of the species *Allolobophora caliginosa*, in numbers equivalent to a normal field population, and kept them at 15°C for a period of forty-five days. The cultures were kept moist and aerated during this time, and their oxygen consumption recorded. The rate of nitrate and ammonium accumulation was measured for a period of fifty days (Table 14).

The cultures with worms consumed 410 litres per g of oxygen more than those without worms, and of this, 200 litres per g was estimated to have been consumed by the earthworms, the remainder being due to increased microfloral activity. Thus, the rate of litter decomposition estimated from the accumulation of nitrate and ammonia was 17–20% greater in the earthworm cultures; half of this was thought to be directly due to the worms and the other half to the stimulation given to other decomposer organisms. Antlett (1951) also showed that *E. foetida* could stimulate microbial decomposition in soil. After five months, the microbial population in soil with earthworms, and to which grape husks had been added, was four to five times greater than in similar soil without earthworms.

8. Earthworms and soil fertility

8.1 Effects of earthworms on soil structure

The activities of earthworms that have most influence on soil structure are: (1) Ingestion of soil, partial breakdown of organic matter, intimate mixing of these fractions, and ejection of this material as surface or sub-surface casts. (2) Burrowing through the soil and bringing subsoil to the surface.

During these processes they thoroughly mix the soil, form water-stable aggregates, aerate the soil and improve its water-holding capacity.

8.1.1 *Breakdown of soil particles*

The presence of smaller amounts of sand and other larger soil fractions in wormcasts than in the surrounding soil, has been taken as evidence that worms can comminute or break down mineral particles to smaller units (Teotia *et al.*, 1950; Joshi and Kelkar, 1952; Shrikhande and Pathak, 1951). Evans (1948) reported that the amounts of coarse sand relative to silt and clay in two pastures with large numbers of earthworms increased with depth, and they suggested that earthworms might be breaking down the coarse sand in the surface soil.

In pot experiments with earthworms, it was shown that the size of particles of granite in the soil was less when earthworms were present, than when there were none (Bassalik, 1913). Similar experiments with soil (Blancke and Giesecke, 1924) and basalt (Meyer, 1943) confirmed these results although the number of particles broken down was small. It seems unlikely however, that the amount of comminution that occurs in this way is either sufficient or rapid enough to be of any importance compared with weathering processes. It is by no means certain that earthworms do in fact comminute the mineral fraction of soils. Nijhawan and Kanwar (1952) reported that earthworm castings contained more coarse fractions than the parent soil, and they suggested that worms reject the finer particles when feeding.

8.1.2 *Turnover of soil*

Large amounts of soil from deeper layers are brought to the surface and deposited as casts. The amounts turned over in this way differ greatly with habitat and geographical region, ranging from 2 to 250

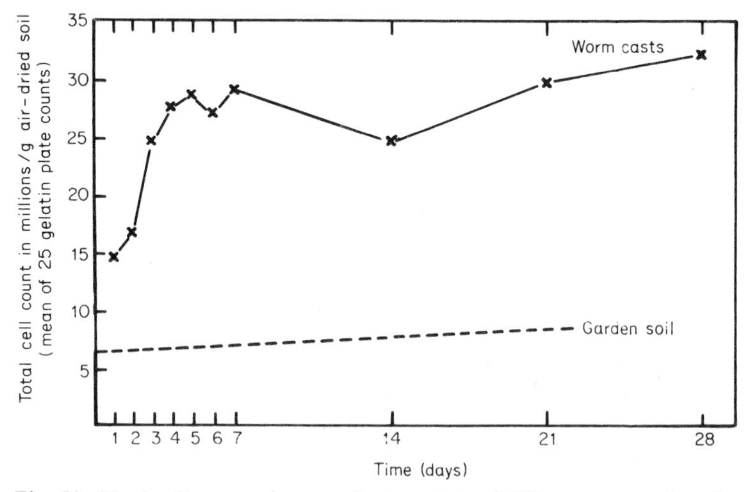

Fig. 51 Total cell counts from earthworm casts of different ages, and garden soils.
(*After Stöckli, 1928*)

tonnes per ha (see Chapter 5). This is equivalent to bringing up layers of soil between 1 mm and 5 cm thick, to the surface every year. In addition, large amounts are deposited either as subsurface casts, or within burrows, so the total soil turnover is even greater. The long-term effect of such turnover is to provide a stone-free layer on the soil surface, particularly in old pasture, which may have such layers 10–15 cm deep.

The great importance of this turnover can be demonstrated by comparing the soil profile of a mor soil containing few worms, with that of a mull that usually has many. The former is distinctly horizontally stratified or layered, whereas the soil horizons in the mull soil blend into each other and are often very difficult to distinguish.

8.1.3 *Formation of aggregates*

Aggregates are mineral granules joined together in such a manner that they can resist wetting, erosion or compaction and remain loose when the soil is either dry or wet. A soil that is rich in aggregates remains well aerated and drained, so that the formation of aggregates is of prime importance to fertility.

Most workers have agreed that earthworm casts contain more water-stable aggregates than the surrounding soil (Bassalik, 1912; Gurianova, 1940; Hopp and Hopkins, 1946; Dawson, 1947; Dutt, 1948; Joachim and Panditesekera, 1948; Chadwick and Bradley, 1948; Swaby, 1949; Bakhtin and Polsky, 1950; Ponomareva, 1950; Teotia *et al.*, 1950; Finck, 1952; Nijhawan and Kanwar, 1952; Mamytov, 1953; Low, 1955; Guild, 1955).

In a typical experiment, the percentage of aggregates in soil to which earthworms were added, was compared with those in soil without earthworms (Hopp, 1946). After three days, the worm soil had 12% of large aggregates whereas the soil without worms contained only 5·9%. It is still not certain how these aggregates are formed. Not all species of worms are equally efficient in producing aggregates, the degree of stability of their casts depending very much on the food and behaviour of the worms (Guild, 1955). For instance, the surface-feeding *Allolobophora longa* and *Lumbricus terrestris* formed the largest aggregates in soil, whereas the non-surface-casting species *Lumbricus rubellus* and *Dendrobaena subrubicunda* formed only a few small aggregates. The addition of succulent material such as alfalfa hay to soil, greatly increased the amount of aggregation that occurred (Dutt, 1948; Dawson, 1948) and the stability of casts may depend very much on the availability of certain types of plant organic matter (Hoeksema *et al.*, 1956). The suggestion was that plant remains that have passed through earthworms reinforce and hold the aggregates together, but Swaby (1949) was unable to produce aggregates experimentally from soil and macerated roots. It is certainly well established that the stability of earthworm casts from land under grass or forest is greater than those under arable crops such as lucerne or cereals (Table 19) (Dutt, 1948; Teotia *et al.*, 1950; Mamytov, 1953.)

An alternative theory is that aggregates are formed by internal secretions, which cement soil particles together as they pass through the intestines of earthworms (Bakhtin and Polsky, 1950), but the process cannot be as simple as this, or casts from arable and grassland soil would not differ in stability as much as they do. Another possible explanation is that the stable aggregates in worm-casts could be produced from soil particles, cemented together by calcium humate that is synthesized in the earthworm intestine from decaying organic matter, and calcium from the calciferous glands. Certainly, calcium humate can stabilize soils (Swaby, 1949) and this could account for the correlation between the stability of worm-casts and their humus content (Ponomareva, 1950).

Some workers have suggested that the stability of aggregates in worm casts and worm-worked soil, is due to bacteria producing stabilizing materials in the casts, because although organic matter causes aggregation, it does so only in soils where micro-organisms are present (Waksman and Martin, 1939), and some bacteria are known to produce secretions such as polysaccharide gums (Geoghegan and Brian, 1948). However, Dawson (1947) could find no direct relationship between the numbers of bacteria in soils and casts, and the proportion of water-stable aggregates in them. When pasture soil was incubated together with fresh worm casts, there was a rapid growth of fungi and an increase in the amount of aggregation (Swaby, 1949). However, this is inconclusive because fungi are usually fewer in newly-deposited earthworm casts (Parle, 1963), but the stability of casts of *A. longa* gradually increased for fifteen days, then began to decrease, and these changes were correlated with the growth of fungal hyphae. It has been reported that the percentage of water-stable aggregates in different soils, greatly increased when they were incubated with plant residues, and again, the greatest degree of aggregation occurred after sixteen days, coinciding with the vigorous development of fungal hyphae. As the hyphae disappeared, the larger aggregates disappeared, and there were more small aggregates; this was attributed to the cementing effect of bacterial gums.

It is by no means certain how long the stability of aggregates in earthworm casts lasts, and how important this is in determining the long-term effects of the aggregation; most of the evidence is that they break down within weeks or months. It seems from the available evidence, that aggregates can be formed by one or more of several of these different methods. For instance, bacterial gums may be important in forming aggregates in earthworms' intestines, and fungi in the casts, and organic matter is required to give the aggregates structure. Earthworms cannot be considered essential for the formation of aggregates, because aggregates occur in soils that contain no earthworms. Probably, aggregates are readily formed in grassland without the intervention of earthworms, but in forest soils worms have a much more important part in the formation of aggregates (Jacks, 1963).

8.1.4 *Aeration, porosity and drainage*

Earthworms improve soil aeration by their burrowing activity, but they also influence the porosity of soils by their effect on soil structure. It has been calculated (Wollny, 1890) that earthworm burrows increase the soil–air volume from 8% to 30% of the total soil volume, but this was in rather unnatural culture conditions. Stöckli (1928) estimated that in a garden soil with 2·4 million earthworms per ha, their burrows occupied 9–67% of the total soil–air space, which can be compared with 38 and 66% for a ley and a pasture respectively (Evans, 1948). Probably, a more realistic estimate is that earthworm burrows constitute only 5% of the total soil volume (Stöckli, 1949). Teotia (1950) claimed that earthworm activity increased the porosity of two soils from 27·5% to 31·6% and 58·8% to 61·8% respectively. Clearly, earthworms greatly increase the aeration and structure of soils.

Soils with earthworms drain from four to ten times faster than soils without earthworms (Guild, 1952; Teotia, 1950; Hopp and Slater, 1949). After twenty-four hours free drainage, there was little difference in moisture content, but soil without worms was waterlogged, whereas worm-worked soil was well aggregated, with water held as capillary water within the larger aggregates (Guild, 1955). Earthworms have increased the field capacity of some New Zealand soils, compared with soils without earthworms, by as much as 17%

(Stockdill and Cossens, 1955). Clearly, earthworms influence the drainage of water from soil and the moisture-holding capacity of soil, both of which are important factors for growing crops.

8.2 Earthworms as indicators of soil type

Several workers have proposed that the species of earthworms that occur in a soil can be indicators of the soil type and its properties (Müller, 1878; Bodenheimer, 1935; Saussey, 1959; Lee, 1959; Volz, 1962). Ghilarov (1956, 1965) is the chief modern proponent of this theory and his work has been comparatively successful. Many other attempts to diagnose soils by the earthworms they contain have been unsuccessful, probably because those ecological factors that favour multiplication of earthworms, such as moisture capacity, pH, organic matter content, etc., are not always those properties directly linked to soil type. Obviously, certain species of earthworms are associated with extreme soil types, but we need a much more thorough knowledge of the distribution of earthworms in different soils, and widespread and very thorough soil analyses, before we can hope to diagnose soils successfully on the basis of their earthworm populations.

8.3 Effects of earthworms on crop yields

It is now certain that earthworms have beneficial effects on soils and many workers have attempted to demonstrate that these effects cause increased yields of crops. Some of the effects of earthworms on soil take much too long to produce detectable effects on plant growth, in experiments lasting only a few months or even years. There is also the difficulty in distinguishing between the effects of living earthworms on soil conditions and nutrient content, and those due to the addition of nitrogenous compounds and other nutrients to soil, from the bodies of worms that die and decay during the course of experiments.

Many workers have obtained inconclusive results from pot experiments because of this error (Wollny, 1890; Chadwick and Bradley, 1948; Baluev, 1950; Joshi and Kelkar, 1952; Nielson, 1953). However, Russell (1910) accounted for release of nutrients from dead worms, by adding the same number of dead worms to his control soils, as he added live worms to test soils. When he added 0·5 g (live weight) of earthworms to each kg of soil, he obtained increased dry matter yields of the order of 25%, and he attributed this to improvements in the physical condition of soils, for instance, there was less evaporation from pots that contained worms, because the soil surface was covered by casts.

Large numbers of earthworms added to soil doubled the dry-matter yield of spring wheat grown in the soil, and increased grass yields four times and clover yields ten times, although pea yields were decreased (van Rhee, 1965). Kahsnitz (1922) claimed that the addition of live worms to a garden soil increased yields of peas or oats by 70%, but the numbers of worms added were very large. Herbage plants grown in poorly-structured soil to which live earthworms were added at a rate of 120 per m^2 (together with organic matter) yielded 3,160 kg per ha, whereas when dead earthworms were added instead, the yield was only 280 kg per ha (Hopp and Slater, 1948, 1949). These workers also tested the influence of four species of earthworms on soil fertility, and concluded that they consistently increased yields of millet, lima beans, soybeans and hay, and that adding live worms increased yields more than adding dead ones. The growth of soybeans and clover in soils with poor structure was stimulated more than that of grass and wheat.

L. terrestris and *L. rubellus* increased yields of barley in heavily manured soils in garden frames (Uhlen, 1953), and winter wheat yielded more in plots to which live worms were added (Dreidax, 1931). Ribaudcourt and Combault (1907) also reported that the addition of worms to small field plots increased yields, but they did not account for the effects of dead worms.

When individuals of *Allolobophora caliginosa* were allowed to feed for eight weeks, in a soil that contained 30 g of dry dung per kg of soil, and were then removed, and rye grass planted in the soil, yields doubled in the earthworm-worked soil (Waters, 1951). This showed conclusively, that worms can increase yields in pot experiments and other workers confirmed this in field experiments.

There is evidence from field experiments in New Zealand, that the addition of European species of earthworms to sown pastures can increase crop yields. The soils were usually acid, so lime was added to counteract this, then colonies of about twenty-five individuals of

A. caliginosa were added, and four years later, around each inoculation point there was a greener and more densely-covered area, several metres in diameter. After eight years, the areas of earthworm activity had spread as far as 100 m from the initial inoculation points (Hamblyn and Dingwall, 1954; Richards, 1955; Stockdill,

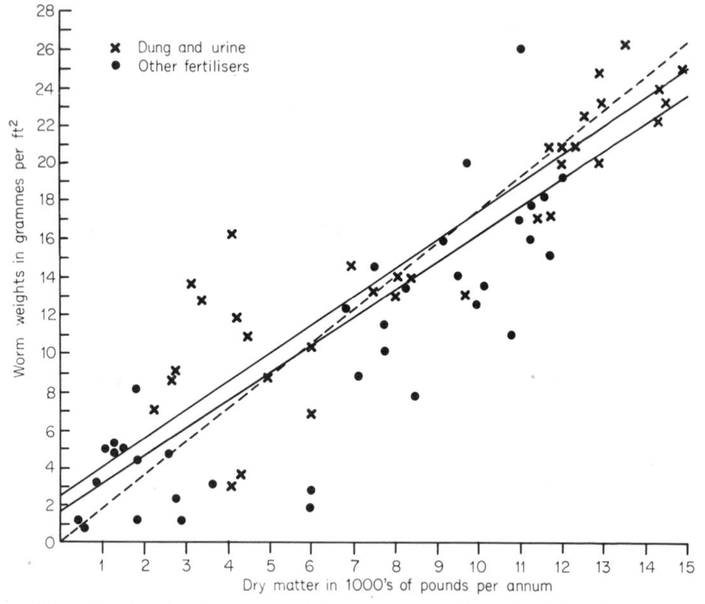

Fig. 52 Relation between weight of worms and productivity of pasture. (*After Waters, 1955*)

1959). There are many instances of increased yields caused by earthworms on soils from different parts of New Zealand, for instance, *A. caliginosa* increased production by between 28% and 100% and yield increases in mixed swards ranged from 28% to 110% (Stockdill and Cossens, 1966; Nielson, 1953), and Waters (1955) reported similar yield increases (Fig. 52).

There have been a few studies on the effect of earthworms on the growth of forest trees. Live earthworms, when added to pots, increased the growth of two-year-old seedlings of oak (*Quercus rober*) by 26%, of green ash (*Fraxinus pennsylvanica*) by 37% (Zrazheyski, 1957) and black spruce seedlings significantly increased in weight when earthworms were added to the soil in which they were grown (Marshall, 1972).

There have been suggestions that earthworms produce plant growth substances, for instance, Nielson (1965) claimed that he detected such substances in eight species of lumbricids and two megascolecids, and stated that they were secreted into the alimentary tract and voided with the faeces. Hopp and Slater (1949) also suggested that earthworms release substances beneficial to plant growth, principally in summer.

All studies of the influence of earthworms on yield of crops have been based on a method using addition of live or dead worms to worm-free soil. Probably a better approach would be to compare the yield in plots with natural populations, with those from which worms have been removed either with electrical stimuli (this might, however, kill worms and leave their carcasses behind), or formalin extraction. Dobson and Lofty (1956) used chemicals to kill earthworms in moorland soils, but their chemicals may have affected other soil animals.

8.4 Soil amelioration by earthworms

It has been proved so conclusively that earthworms aid soil fertility, that there have been many attempts to add earthworms to poor soil or to encourage the build up of populations by addition of organic matter or fertilizers. Farms that breed earthworms for adding to poor land have been set up in both Europe and the United States and some of these can produce as many as half a million worms per day. Such addition of earthworms to soil seems particularly promising in reclaiming flooded areas that are subsequently drained and put into cultivation, as in the Dutch polders (van Rhee, 1969, 1972). For instance, earthworms (*A. caliginosa* and *L. terrestris*) were introduced to polder soil in which fruit trees were grown, at a rate of about 800 worms per tree. More roots grew in the worm-inoculated soils than in those without worms (van Rhee, 1969, 1971). Worms multiplied rapidly in polder soils, for instance, *A. caliginosa* increased from 4,664 individuals to 384,740 in three to four years, and *Allolobophora chlorotica* from 2,588 to 12,666 in the same time (van Rhee, 1969).

Earthworms have been successfully introduced to newly-established areas of artesian irrigation in Uzbekistan, U.S.S.R., in order to improve soil formation (Ghilarov and Mamajev, 1966). Sometimes, however, amelioration projects are foredoomed because no attention is paid to introducing suitable species of worms. Not all worm species are active in soil formation, and often those species that can be most easily bred are the most useless in reclamation, for example, ***Eisenia foetida*** is easy to breed and is commonly sold by breeders, but this species is really a manure or compost species and

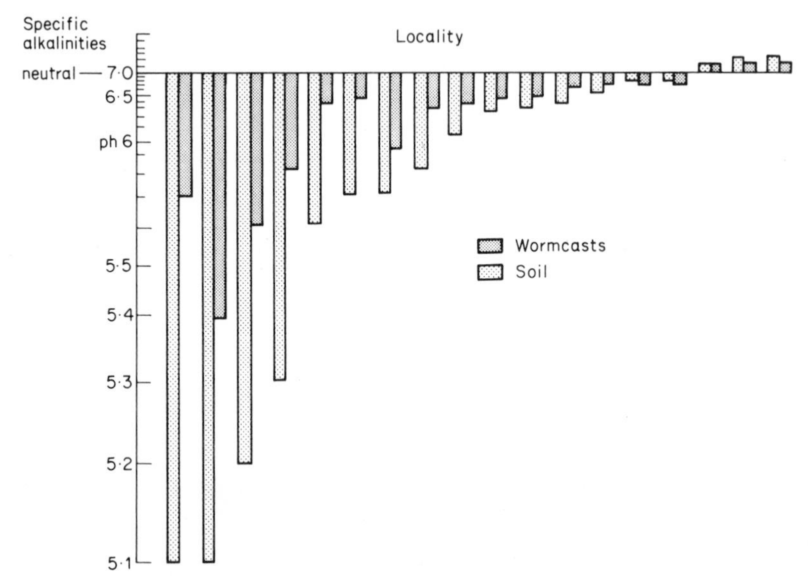

Fig. 53 Average pH of the soil and wormcasts from eighteen localities. (*Adapted from Salisbury, 1923*)

cannot survive long or thrive in field soils (Grant, 1955). Any increases in yield after adding *E. foetida* to soil are short-lived, and are due to the decomposition of dead worms rather than to worm activity. Any given area of soil can support a certain limited earthworm population, and adding more than this will not increase fertility unless more organic food is supplied and soil conditions improved.

Adding earthworm casts to soil can improve greatly its structure and fertility. Casts usually have a higher pH (Fig. 53), and more total and nitrate nitrogen, organic matter, total and exchangeable magnesium, available phosphorus, base capacity, and moisture equivalent (Lunt and Jacobson, 1944) (Table 19). Adding lime to soil,

TABLE 19

The stability of earthworm casts and the surface soil

Source of sample	Percentage aggregates					
	Surface soil (0–7·5 cm)			Earthworm casts		
	Sample a	Sample b	Mean	Sample a	Sample b	Mean
Cultivated (silt loam)	5·8	7·4	6·6	19·0	19·6	19·3
Pasture (3 years)	43·4	48·0	45·7	56·6	59·6	58·1
Forest	58·2	59·0	58·6	65·6	70·8	68·2

(*From Dutt, 1948*)

usually increases earthworm populations, for instance, the addition of 1 ton of lime per acre to New Zealand soils caused an increase of 50% in numbers of *A. caliginosa* (Stockdill and Cossens, 1966). This process has been accelerated by inoculating with more worms in spring when soils are moist, after adding lime, and it was found that worms then spread from an inoculation point at a rate of about twelve metres per year.

9. Effects of agriculture on earthworm populations

9.1 Effects of cultivations

It is now well established that grassland usually contains more earthworms than arable land (Tables 5, 7 and 10). This could be due to mechanical damage during cultivation, to the loss of the insulating layer of vegetation, or to a decreased supply of food as the organic matter content gradually decreases. Many workers have considered that these differences are largely due to mechanical damage during cultivation. When old grassland is ploughed, the number of earthworms in it steadily decreases with time after ploughing (Graff, 1953). In one such study (Evans and Guild, 1948), five years after grass was ploughed, the earthworm population had declined by 70%, although the population was unchanged by the first ploughing of the sward, so it is unlikely that mechanical damage was a primary cause of the decreased numbers of worms. Hopp and Hopkins (1946) also reported that cultivation of arable land in late spring did not decrease earthworm numbers. Indeed, it would be surprising if mechanical damage by ploughing was very important, because the plough merely turns the soil over, and probably has little effect on worms with deep burrows. Preparation of seed beds by rotary cultivation, harrowing, disking or rolling can be expected to damage more earthworms, but the regenerative powers of earthworms are so great that only a few would be killed outright. Edwards and Lofty (1971) investigated the effects of maximal and minimal cultivation of grass plots, on earthworm populations. They compared plots ploughed and cultivated in spring with others that were unploughed. The more the soil was cultivated during the first two seasons, the greater was the number and weight of earthworms in the soil (Table 20). Clearly, mechanical damage did not decrease populations in this experiment. Zicsi (1958) stated that cultivation after harvest had drastic effects on earthworm populations, with a mortality of 16·1% after stubble-stripping, 39·3% after summer ploughing, and 67·2% after cultivating further, but these were only

TABLE 20

Earthworm populations in cultivated plots

Treatment	Earthworm populations			
	wt/g m²	% control	no./m²	% control
Control (unploughed)	6·28	100	4·18	100
Minimal cultivation (once spring ploughed)	6·35	101	5·06	121
Minimal cultivation (twice spring ploughed)	6·50	104	4·37	104
Maximal cultivation (once spring ploughed)	8·13	129	5·30	127
Maximal cultivation (twice spring ploughed)	7·35	117	5·55	133
Maximal cultivation (once autumn ploughed)	8·66	138	6·39	153
Maximal cultivation (once autumn ploughed)	10·30	164	6·88	164

(*From Edwards and Lofty, 1970*)

surface-living earthworms. In later work (Zicsi, 1969) he reached the conclusion that moderate cultivation, e.g. disk cultivation, favoured earthworms by loosening the soil, and he stated that by careful selection of cultivating machinery, earthworm numbers can be increased.

Workers in the United States have postulated that decreased numbers of earthworms in arable land could be prevented by leaving the remains of the crop lying on the soil to form an insulating layer (Hopp, 1946; Hopp and Hopkins, 1946; Slater and Hopp, 1947). The evidence they presented was fairly conclusive, but winters are usually very cold in the northern United States, and it is unlikely that these conclusions are valid for more temperate areas.

It now seems probable, that the most important factor controlling earthworm populations in arable land, is the amount of organic matter that is available as food for earthworms. It has been shown conclusively, that the availability of food can limit numbers of earthworms in grassland (Satchell, 1955; Waters, 1955; Jefferson, 1956) and arable land (Lofty, unpublished) (Table 10), and Evans and Guild (1948) attributed the decrease of numbers of worms in arable land to the gradual decrease in the availability of food. Quite large populations of earthworms can be maintained in arable land if farmyard manure is regularly applied, but exhaustive cropping without adding organic manure usually decreases earthworm populations to a very low level (Morris, 1922; Evans, 1947; Tischler, 1955; Barley, 1961).

Stubble-mulch farming consists of leaving crop residues on the soil surface and working the soil underneath this layer with subsurface tillage implements called 'Subtillers'. Teotia *et al.* (1950) reported that there were three to five times more earthworms in sub-tilled and stubble-mulched plots than in ploughed plots and this was confirmed by McCalla (1953). Graff (1969) pointed out that mulching favours the deposition of casts on the surface of arable land.

There seems to be little doubt that grass favours earthworm populations, and that the best way of maintaining large earthworm populations in agricultural land, is by ley farming, or at least by the inclusion of leys in the rotation.

9.2 Effects of cropping

It was clear from the discussion in the last section that cropping influences numbers of earthworms in arable and grassland. Hopp and Hopkins (1946) reported that alfalfa-grass plots contained more earthworms than lespedeza-grass plots, and grass plots in orchards contained more than plots with timothy grass. When land was cropped on a rotation that included a pasture ley, the numbers of earthworms changed according to the phase of the rotation. Ponomareva (1950) measured the numbers of earthworms during each phase of a wheat, rye and two-year grass rotation in the U.S.S.R., and reported the largest numbers in the second year of the ley. In a similar experiment in Australia, with a fallow, wheat and two-year grass rotation, Barley (1959) reported that the weight of earthworms after the fallow was only 25% of that occurring in permanent pasture, but by the end of the ley, the weight was at least 70% of that in permanent pasture.

Hopp (1946) reported that earthworm populations in the United States differed greatly under different crops, being least under row

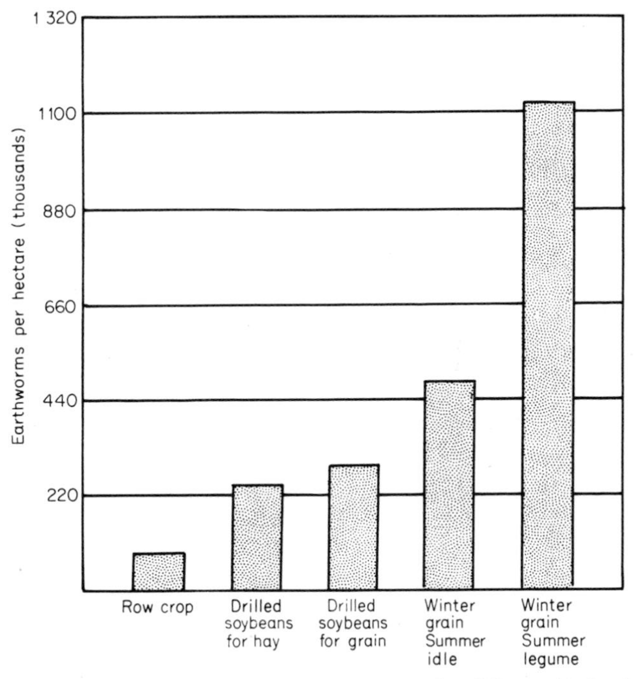

Fig. 54 Earthworm counts in February 1946 under different kinds of annual cultures.
(*After Hopp, 1946*)

crops and greatest in plots growing winter cereal and summer legumes (Fig. 54). The more often row crops were grown, the smaller were the earthworm populations (Fig. 55). This was substantiated in other studies (Hopp and Hopkins, 1946) in which the smallest average earthworm populations occurred under continuous maize; there were rather larger populations under continuous soybeans, much bigger ones under continuous winter cereals, and earthworm numbers were as large as those in pasture, in soil growing winter cereals followed by legume hay. Probably, one of the more important factors affecting the influence of crops on earth-

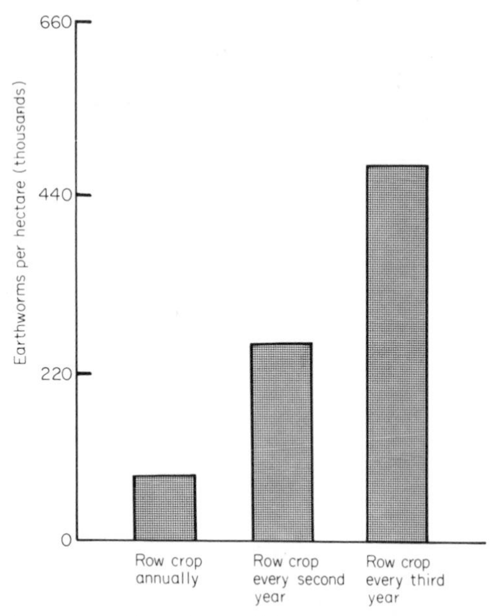

Fig. 55 Earthworm count in February 1946, after row cropping in 1945, in plots planted to row crops annually, every second year, and every third year. (*After Hopp, 1946*)

worm populations is the proportion of the plant material that is returned to the soil after harvest.

9.3 The effects of fertilizers

It has been clearly established that organic fertilizers affect numbers of earthworms in soil (see Chapter 5), but there have been fewer studies of the influence of inorganic fertilizers on earthworm populations. The effects of fertilizers on earthworms may be direct, for instance, by changing the acidity of soil, or indirect, by changing the form and quantity of the vegetation that ultimately provides food for worms. For instance, application of superphosphate and lime to a pasture, caused a dense clover sward to develop and this in turn increased the weight of earthworms in the soil about four-fold (Johnstone-Wallace, 1937).

There is good evidence that nitrogenous fertilizers favour the build-up of large numbers of earthworms. Large amounts of nitro-chalk applied to pastures, indirectly increased the earthworm population due to greatly increased grass production (Watkin, 1954). Jacob and Wiegland (1952) also reported increased numbers of earthworms, after application of different forms of nitrogenous fertilizers; they estimated that there were 128 earthworms per m^2 in

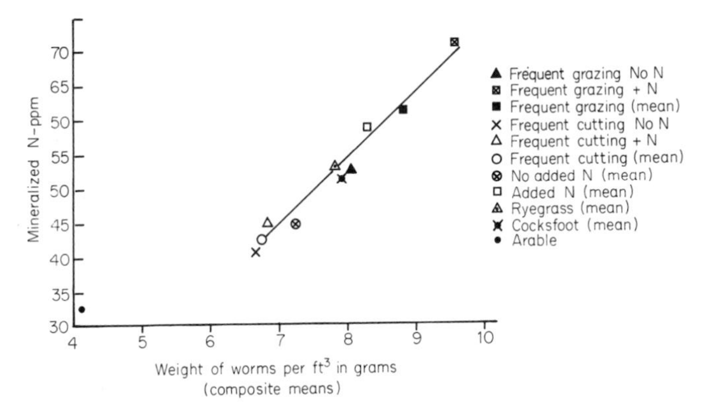

Fig. 56 Relationship between weight of earthworm populations and amount of nitrogen mineralized on incubation of soil samples.
(*After Heath, 1962*)

plots without nitrogen, and 176 per m^2 in plots to which nitrogenous fertilizers had been added. Heath (1962) reported a linear correlation between the amounts of nitrogenous fertilizers used on leys and the weight of earthworms in the soil (Fig. 56). Applications of nitro-chalk and nitrate of soda also resulted in increased earthworm populations (Escritt and Arthur, 1948), but on Park Grass, Rothamsted, smaller numbers of worms occurred in plots that had received annual doses of nitrate of soda, than in unmanured plots (Richardson, 1938).

Lime also seems beneficial to earthworms (Richardson, 1938; Crompton, 1953; Jefferson, 1955), and this is probably because most species of earthworms tend to avoid acidic soils (see Chapter 5).

Probably, lime is only effective in soils that have a pH below 4·5–5·0, because above this level, acidity does not greatly affect populations of many species of earthworms. Crompton (1953) believed that the way lime influenced earthworm populations, was by helping to decompose organic matter into a form that earthworms could eat, but this conclusion is dubious, if earthworms are to be considered as primary litter breakdown organisms.

It is not clear what the effects of superphosphate on earthworms are, and they probably differ with the soil conditions (Bachelier, 1963). Doerell (1950) concluded that superphosphate was very beneficial to earthworms, but Escritt and Arthur (1948) reported that this fertilizer decreased numbers of earthworms in grass plots.

Most other mineral fertilizers have little effect on earthworm populations, for instance, Uhlen (1953) could find no influence of commercial fertilizers on earthworms. Small increases in numbers of worms, in plots treated, with mineral fertilizers, have been occasionally reported (Ogg and Nicol, 1945; Jacob and Wiegand, 1952).

TABLE 21

Wormcasts on grassland (Park Grass, Rothamsted)

Treatment	Number of casts (thousands per acre)	No./m^2
Sulphate of ammonia + minerals	0	0
Sulphate of ammonia + minerals + lime	127	31
Nitrate of soda + minerals	153	38
Nitrate of soda + minerals + lime	161	40
Unmanured	245	60
Unmanured	294	72
Unmanured + lime	276	68
Dung	423	104
Dung + lime	337	83

(*From Richardson, 1938*)

There is good evidence that sulphate of ammonia is antagonistic to earthworm populations (Escritt and Arthur, 1948; Rodale, 1948; Jefferson, 1955). Slater (1954) stated that the use of sulphate of ammonia on grass plots, for three years in succession, decimated the earthworm population, and Richardson (1938) reported that annual treatment with sulphate of ammonia, completely eliminated earthworms from plots on Park Grass at Rothamsted (Table 21). The reason that sulphate of ammonia is unfavourable to earthworms is probably that it makes soils more acid, its effect being greatest in soils that are already rather acid; this was certainly so on Park Grass. Sulphate of ammonia is so drastic in its effect that Slater (1954) suggested it could be used as a means of eliminating earthworms from golf courses.

9.4 Mortality from pesticides

Large amounts of insecticides, herbicides and fungicides are applied to soil to control pests. Some of these are general biocides and may also kill earthworms. Arsenic compounds, especially lead arsenate (Escritt, 1955; Polivka, 1951) and copper sulphate, which were used as pesticides before the Second World War, are toxic to earthworms when they occur in the large amounts that sometimes accumulate in orchards and some cultivated soils. For instance, it has been shown that copper fungicides are toxic to earthworms (Duddington, 1961), and Raw and Lofty (1959) reported that orchard soils that contained large copper residues had few earthworms, and a thick mat of undecayed organic matter on the soil surface (Plate 8a).

Fumigants such as D-D, metham sodium and methyl bromide, used to control soil pathogens and nematodes, permeate the soil as vapours and kill most of the worms, even those that live in burrows deep in the soil (Buahin and Edwards, 1963). Chloropicrin was also very toxic to all earthworms (Blankwaardt and van der Drift, 1961).

Herbicides can affect earthworm populations either directly, or indirectly by killing the vegetation on which the worms feed. Not many herbicides decrease earthworm populations directly. Fox (1964) reported that TCA decreased earthworm populations and Wojewodin (1958) confirmed this, although the dose he tested was very large (100 kg per ha). DNOC, chlorpropham and propham had slight effects on earthworms in plot tests (Bauer, 1964). The triazine herbicides tend to kill a few earthworms, for instance, atrazine slightly decreased their numbers (Fox, 1964). Edwards (1970) reported that simazine had the same effect and although he reported that another triazine, Shell 19805 ('Bladex'), significantly increased numbers of earthworms compared with untreated plots, the latter effect was probably an indirect one due to greater amounts of dead plant material available as food for worms. More earthworms occurred in soil that grew successive crops of wheat after treatment with the herbicide paraquat, than in similarly cropped plots that were ploughed (Table 22) (Edwards and Lofty, 1971). This was probably due to the cultivations and changes in available organic matter, rather than a direct effect of paraquat.

TABLE 22

A comparison of earthworms in ploughed plots and paraquat-treated slit-seeded plots

	No. of worms (per m^2)		Wt of worms (gm/m^2)	
Year	Paraquat treated	Ploughed	Paraquat treated	Ploughed
1967	431	25	70·9	2·0
1968	431	331	108·5	43·7
1969	344	487	36·0	25·2

(*From Edwards and Lofty, 1970*)

There have been many studies of the effects of insecticides on earthworms, many of which were reviewed by Davey (1963).

Aldrin seems to have little effect on numbers of earthworms at normal doses (Edwards and Dennis, 1960, 1967; Hopkins and Kirk, 1957; Bigger and Decker, 1966; Legg, 1968), although at very high doses several workers have reported decreased populations (Polivka, 1953; Schread, 1952; Hopkins and Kirk, 1957; Heungens, 1969).

BHC is not very toxic to earthworms at normal rates of application (Polivka, 1953; Gunthart, 1947; Morrison, 1950; Grigor'eva, 1952; Richter, 1953; Weber, 1953; Lipa, 1958; Ghilarov and Byzova, 1961), although it has sometimes killed worms at higher rates (Schread, 1952; Hoy, 1955). Lipa (1958) and Bauer (1964) reported increased numbers of earthworms after treatment with BHC.

Chlordane is extremely toxic to earthworms (Schread, 1952; Polivka, 1953; Hopkins and Kirk, 1957; Doane, 1962; Edwards,

1965; Lidgate, 1966; Long *et al*, 1967; Legg, 1968), so much so, that it is the chemical most commonly used for earthworm control on sports fields.

Many workers have studied the effects of DDT on earthworm populations, and most have concluded that normal rates of application do not harm earthworms (Fleming and Hadley, 1945; Goffart, 1949; Fleming and Hadley, 1950; Polivka, 1951, 1953; Richter, 1953; Hopkins and Kirk, 1957; Barker, 1958; Edwards and Dennis, 1960, 1967; Ghilarov and Byzova, 1961; Stringer and Pickard, 1963; Edwards, 1965; Doane, 1962; Thompson, 1971). Other workers have reported that DDT did affect earthworms, for instance Rodale (1948) stated that DDT was 'instant death' to earthworms, but supplied no supporting data, and Greenwood (1945) reached the same conclusion but based his results on only one pot experiment using eight worms. Hoy (1955) found some effects of large doses in field tests.

Dieldrin is related to aldrin, and its toxicity to earthworms is similar (Polivka, 1953; Hopkins and Kirk, 1957; Luckman and Decker, 1960). Although Schread (1952), Doane (1962) and Legg (1963) reported that large doses killed worms, there is little doubt that the amounts normally applied to agricultural land do not kill earthworms.

Endrin had no effect on earthworms at 6·2 kg per ha (Hopkins and Kirk, 1957), although larger doses did. Patel (1960) reported that earthworms were susceptible to endrin, and Edwards and Lofty (1971) and Thompson (1971) also reported that normal agricultural doses of endrin decreased earthworm populations, so it seems that endrin is a chemical that is relatively toxic to earthworms.

Heptachlor is related to, and is almost as toxic to earthworms as chlordane. Polivka (1953) reported that all the doses of heptachlor he tested decreased earthworm populations, and this has been confirmed by Rhoades (1963) and Edwards and Arnold (1966). Telodrin did not affect numbers of earthworms in studies by Edwards (1965), but Kelsey and Arlidge (1969) reported that it did. Its effects are probably not drastic because its toxicity to invertebrates resembles that of aldrin.

All the chemicals discussed so far are organochlorines but several workers have also investigated the effects of organophosphate insecticides on earthworm populations. Azinphosmethyl did not affect earthworm populations (Hopkins and Kirk, 1957), but carbofuran did (Kring, 1969; Thompson, 1971). Chlorfenvinphos had slight effects (Edwards *et al.*, 1967), but diazinon had none and disulfoton was only slightly toxic to worms (Edwards *et al.*, 1967). Dyfonate had a slight effect on numbers of earthworms (Edwards, 1970) and, so did malathion (Voronova, 1966), 'Dursban' (Whitney, 1967; Thompson, 1971), and fenitrothion (Griffiths *et al.*, 1967). Neither 'Guthion' (Hopkins and Kirk, 1957), malathion (Edwards, unpublished; Hopkins and Kirk, 1957) nor menazon (Way and Scopes, 1968; Raw, 1965) affected earthworms. Parathion has been reported as moderately toxic to earthworms, particularly in large doses (Goffart, 1949; Schread, 1952; Weber, 1953; Scott, 1960; van der Drift, 1963; Heungens, 1966), but Hyche (1956) and Edwards *et al.* (1967) did not find that this chemical decreased earthworm numbers greatly.

Phorate is extremely toxic to earthworms (Edwards *et al.*, 1967; Way and Scopes, 1965, 1968; Kelsey and Arlidge, 1968) and may almost eliminate earthworms from soil even at normal agricultural rates. Sumithion and trichlorphon have little effect on earthworms, although some are killed by zinophos (Edwards, unpublished).

Only three carbamate insecticides have been tested for their effects on numbers of earthworms and they all kill earthworms very readily. Carbaryl is toxic to earthworms, even at very low doses, and causes ulcers to form on their body surfaces (an der Laan and Aspock, 1962; Heungens, 1966; Edwards, 1965; Legg, 1968; Thompson, 1971). Aldicarb also decreased numbers of earthworms (Edwards *et al.*, 1971), and so did carbofuran (Kring, 1969; Thompson, 1971).

9.5. Uptake of pesticides into earthworms

Earthworms are not very susceptible to pesticides and can live in soil containing large amounts of some persistent insecticides. These persistent organochlorine insecticides are lipophilic, and gradually become absorbed from soil into the earthworm tissues, as the worms pass soil through their intestines. This may be very important ecologically, because earthworms are eaten by many species of birds and several species of mammals, and these animals can further concentrate the pesticide even more.

Barker (1958) reported that he had found large amounts of DDT and its residues (mainly DDE), in the tissues of earthworms taken from soil under elm trees that had been sprayed with DDT to control the insect vectors of Dutch Elm disease, the largest amounts occurring in the crop and gizzard of the worms. Large residues of DDT in earthworms were also reported by Doane (1962) and subsequently many other workers have found such residues in earthworm tissues, (Stringer and Pickard, 1963; Hunt, 1965; Cramp and Conder, 1965; Dustman and Stickel, 1966; Davis and Harrison, 1966; Davis, 1968; Edwards, 1970).

Wheatley and Hardman (1968) investigated the relationship between the amounts of chlorinated hydrocarbon insecticide residues in the tissues of earthworms and the amounts in the soil in which they were living. They found that not all species of earthworms concentrated insecticides into their tissues to the same degree, and that the greatest concentration were found in *Allolobophora chlorotica*, a small species which lives in the surface soil. Smaller concentrations occurred in the larger species (*Lumbricus terrestris*, *Allolobophora longa* and *Octolasium cyaneum*), than in the smaller species (*Allolobophora caliginosa*, *Allolobophora rosea* and *A. chlorotica*), but probably the amounts are correlated more with the habits of these species than their size. The smaller species tend to live mainly in the upper few inches of soil, where insecticides are most likely to be concentrated, both in arable and orchard sites. By contrast, *L. terrestris* lives in well-defined burrows, often 1·2 to 1·8 metres deep, and feeds mainly on surface debris. When large concentrations of residues are on the soil surface, the residues in *L. terrestris* have approximated to those found in other species.

There is evidence (Wheatley and Hardman, 1968), that the amounts of residues in the soil and those in the worms are not linearly related, and there is proportionately less concentration of insecticides into the tissues of worms in soils containing large quantities of residues, than in those with small amounts. In their studies, the degree of concentration changed from about five to ten-fold, when residues in the soil were between 0·001 and 0·01 ppm, to less than unity when the concentrations exceeded 10 ppm in the soil.

However, these conclusions are not supported by the results of an extensive survey in the United States by Gish (1970), who collected

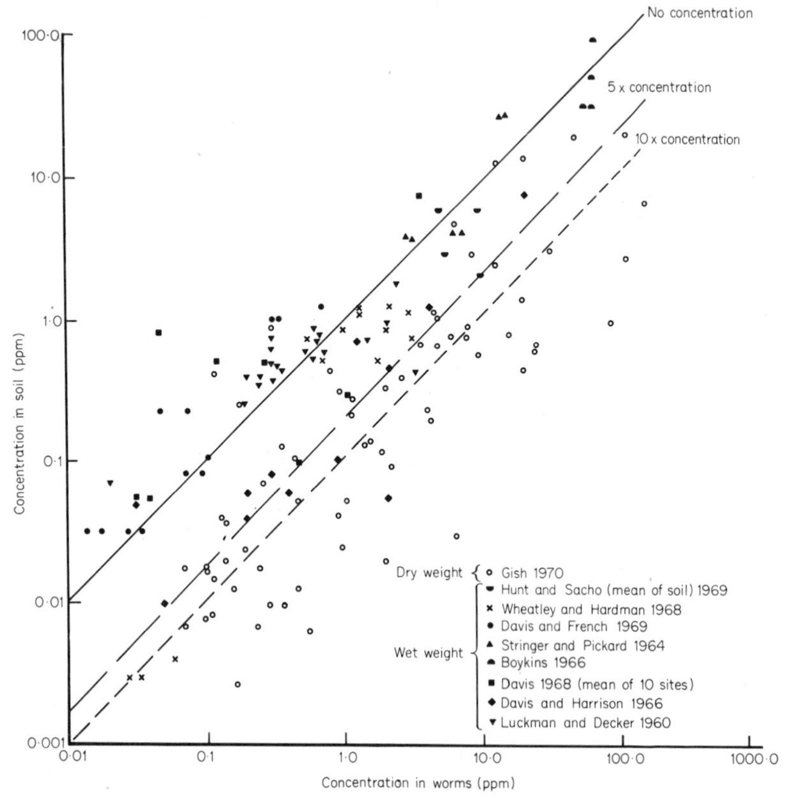

Fig. 57 Concentration of chlorinated hydrocarbon residues from soil to earthworms.
(*After Edwards and Thompson, 1972*)

soils and earthworms from sixty-seven agricultural fields in eight States, and analysed them for organochlorine residues. His data, which are plotted in Fig. 57 together with all other available data, show a more or less linear relationship between amounts in the soil and amounts in the worms, with an average concentration factor of about nine times for all insecticides and doses.

In Gish's studies, the mean amount of total organochlorine residues in the soils was 1·5 ppm and in the worms 13·8 ppm. Residues ranged from a trace to 19·1 ppm in soils, and from a trace to 159·4 ppm in the worms; all of the soils and worms sampled contained some organochlorine residues. Out of the 67 soils, one had residues of two insecticides, 22 had three, 16 had four, 24 had five, 3 had six and one had seven; the same insecticides or their degradation products were present in the worms, as in the soils from which they came. The largest amounts of residues in both soils and worms were in samples from cotton fields and orchards, whereas pastures contained the smallest quantities. One of the most important results of this investigation, was to confirm that worms do not concentrate all organochlorine insecticides from soil to the same degree. The concentration factors for dieldrin and DDT (and its breakdown materials) ranged from 9·0 to 10·6, whereas for aldrin it was 3·3, for endrin 3·6, heptachlor 3·0 and chlordane 4·0. These conclusions are confirmed by Wheatley and Hardman's results, and also by data given by Edwards (1970), except that the concentration factors reported by Gish tend to be somewhat higher than those found by other workers (see Fig. 57). Probably, this is because most workers calculate pesticide residues in soils in terms of dry weight, and in worms on the basis of live weight or wet weight, whereas Gish expressed his data in terms of dry weight of worms. Some of the organochlorine insecticides are broken down to other materials in worms, so that those in aldrin-treated soil can also contain dieldrin residues, in heptachlor-treated soil, heptachlor epoxide residues, and in DDT-treated soil, DDE residues (Smith and Glasgow, 1965; Wheatley and Hardman, 1968; Gish, 1970). All the studies reported have consisted of sampling soils for earthworms and determining the residues they contain. However, this does not show how rapidly worms accumulate organochlorine insecticides, nor how rapidly they excrete it in clean soil. Edwards *et al.* (1971) studied the rate of uptake of DDT into *L. terrestris* in laboratory cultures, and reported that it took nine weeks for the worms to contain 1 ppm, which was the same amount as the soil contained. At this time, the amounts in the worms were still increasing, but two-thirds had been transformed from DDT to DDE. When worms containing DDT residues (DDT + DDE) were put into clean soil they excreted the DDT within three weeks, but the DDE was much more persistent.

There is little evidence that earthworms can concentrate organophosphorus insecticides or any other pesticides into their tissues (Edwards *et al.*, 1967).

9.6 Radioisotopes and earthworms

There has been considerable pollution of soil by radioactive fall-out, and it has been suggested that the dispersal of radioactive contaminants in soil is accelerated by the ability of living organisms to accumulate isotopes in their tissues. There is evidence (Peredel'sky *et al.*, 1957, 1960), that worms were important in dispersing radioisotopes (^{60}Co) through soil.

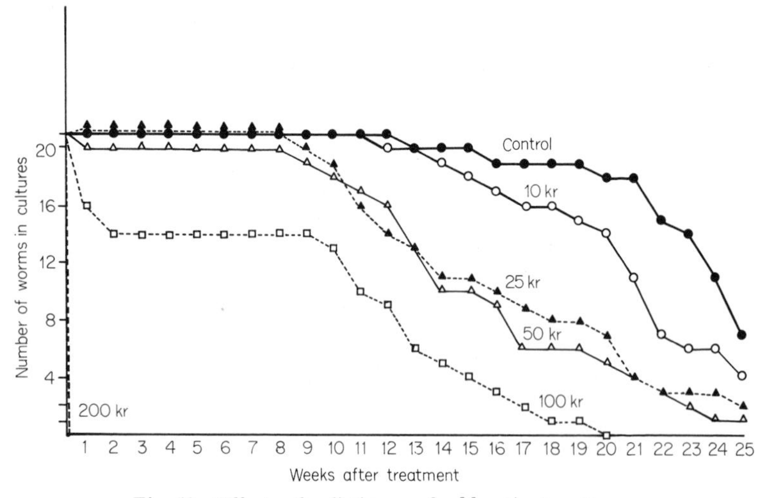

Fig. 58 Effects of radiation on *A. chlorotica* in cultures. (*After Lofty, 1972*)

There has only been one study of the effects of radiation on earthworms; Lofty (1972) studied the effects of treating individuals of *A. chlorotica* with doses of 5, 10, 25, 50, 100 and 200 Kr of gamma radiation (Fig. 58). After treatment with 200 Kr all worms died within a few hours but all other doses except 100 Kr caused little mortality for three months, although thereafter, the worms began to die quite rapidly. All worms treated with 100 Kr were dead within twenty weeks of irradiation. Some of the worms treated with the other doses were still alive after twenty-five weeks.

Crossley *et al.* (1972) studied the uptake of ^{137}Cs by *L. terrestris* and *Octolasium* sp., and reported that about 12% of the amount consumed was assimilated. Maldague and Couture (1971) made similar studies of the uptake of ^{59}Fe by *L. terrestris*, with remarkably good agreement with the data of Crossley *et al.* (1972). Both groups of workers reported that these isotopes did not persist long in the bodies of earthworms, so the movement of isotopes through the soil by the agency of earthworms seems more likely to be a form of passive transport with the soil in the gut.

10. Earthworms as pests and benefactors

10.1 As pests of crops

Certain of the habits of earthworms make them potential pests of crops. The habit of certain species of seizing fallen leaves and pulling them down through the mouth of their burrows, means that if they seize the leaves of growing plants in this way, they damage the plant, sometimes sufficiently to kill it. Zicsi (1954) reported such damage to various crops, and Edwards (unpublished) noted that earthworms destroyed a large part of a lettuce crop in this manner.

As earthworms tunnel through the surface soil they damage small and delicate seedlings by uprooting them (Walton, 1928), and they sometimes damage the roots of delicate plants in pots or flower beds, by tunnelling through their root systems. Such damage is becoming much more important in agriculture with the advent of precision-drilling of crops, with seeds drilled at optimal spacing, and loss or damage cannot be compensated during thinning. Valuable and delicate crops such as tobacco have been reported as being damaged by earthworms in Bulgaria (Trifonov, 1957) and India (Patel and Patel, 1959, Puttaridriah and Sastry, 1961) and rice has been injured in the Phillipines (Otanes and Sison, 1947), in China (Chen and Lui, 1963) and in Japan (Inoue and Kondo, 1962). Damage to vegetable crops was reported by Puttaridriah and Sastry (1961).

There have been reports of worm casts being deposited to such a height in cereal stooks that it was difficult to obtain clean grain when the crop was threshed. Damage to crops that can be indirectly attributed to earthworms, occurs when moles tunnel through arable land in search of earthworms for food.

10.2 As pests of grassland and turf

Grassland usually has very large populations of earthworms, and when these are species which excrete casts on to the surface, they can often be serious pests. They ruin the appearance of ornamental lawns, and the casts are intolerable on the surface of sports grounds, because the projections render the turf unsuitable for golf or bowling. Schread (1952) estimated that as much as 18 tons of casts may be brought up to the surface of golf greens annually. Many thousands of pounds are spent annually in keeping earthworm populations under control in such conditions.

10.3 Control of earthworms

The species of earthworms that are considered to be pests of crops and grass are usually those species that burrow near the surface, and cast on to the surface or come on to the soil surface at night to feed. Hence, a suitable pesticide sprayed on to the soil surface is usually a suitable control agent, and several chemicals originally developed as insecticides, effectively control earthworms (see Chapter 9). The most commonly used of these is the persistent insecticide chlordane, which is applied at a rate of 12 kg active ingredient of dust per ha. Heptachlor and endrin are other organochlorine insecticides that also control earthworms. Unfortunately, all of these chemicals are very persistent in soil, are lipophilic, and hence become concentrated from soil into the bodies of earthworms, and can pass into the tissues of vertebrate predators such as birds and moles when these feed on earthworms. For this reason, less persistent chemicals are preferable, and two that are currently available are carbaryl and phorate, both of which give good control of earthworms at doses of between 6 and 12 kg per ha. However, because they are less persistent than the organochlorine insecticides, more frequent treatments are necessary.

10.4 Transmission of diseases

Earthworms have been incriminated in transmitting many parasites and diseases of animals and plants. They spread soil fungi, including pathogens, throughout the soil (Hutchinson and Kamel, 1956). Spores of many species of fungi can pass through the earthworm intestine's without harm, for instance, there is circumstantial evidence that dwarf bunt (*Tilletia controversa*) may be spread by earthworms, because the teliospores produced by this pathogen are ingested and do not lose their viability during passage through the earthworm's intestine (Hoffman and Purdy, 1964). The pathogenic fungi, *Fusarium* and *Pythium* can also be transmitted through soil by earthworms (Khambatta and Bhatt, 1957; Baweja, 1939).

It has been demonstrated that earthworms greatly affect the infectivity of cysts of the potato root eelworm (*Heterodera rostochiensis* (Ellenby, 1945). Different numbers of earthworms (20, 40, 60 and 100 individuals of *Allolobophora longa* per 25-cm diameter plant pot) were added to soil heavily infected with potato root eelworm cysts. The more earthworms that were introduced to a pot, the more eggs in the cysts hatched, the more rapidly they hatched, and the more viable larvae each cyst produced. He also collected worm-casts and examined the infectivity of cysts from these, as compared with that of cysts from the soil in which the worms lived. The number of larvae produced by fifty-two cysts that had passed through earthworms was 1,974, and the number from seventy-one cysts that had not was only 1,148.

TABLE 23

Numbers of potato root eelworm larvae emerging from cysts ingested by earthworms

No. of earthworms per pot	100	60	40	20	0 (control)
Mean no. of larvae emerging per cyst	158·9	85·4	83·3	71·6	51·8
Standard error	±16·6	±19·7	±22·0	±18·3	±13·8

(*From Ellenby, 1945*)

Ellenby explained that there were three kinds of cysts, those that would have hatched without passing through an earthworm, those that would not hatch unless they have passed through an earthworm, and those that would not have hatched anyway. He was unable to prove conclusively how passage through the worm affected hatchability of cysts, but suggested that probably some digestive enzyme was responsible, although it is also possible that some substance in the casts influenced the hatching of cysts.

Earthworms have been reported as vectors of an animal virus, the foot and mouth disease of domestic animals (Dhennin *et al.*, 1963). Many animal parasites are transmitted from host to host by earthworms, which may be either essential intermediate hosts or merely reservoir hosts to the parasites, transmitting them without any direct influence on the parasites life cycles.

Earthworms are ***essential intermediate hosts*** to a number of tapeworms (Cestoda) and nematodes (Table 24), parasitic in birds and mammals. For the four species of tapeworms listed, the invasive phase that occurs in earthworms is the cysticercus; without such a stage in the earthworm, the life cycle of the tapeworm can proceed no further. Some parasitic nematodes have their larval stages in the bodies of earthworms, and many of these cannot develop to maturity until they reach their final host. For instance, the lung worms (*Metastrongylus*) can only develop to an invasive larval stage in earthworms when they are eaten by pigs, and they can then develop to maturity in the pig's lungs.

Some parasites are transmitted by earthworms without the stage in the earthworm being essential for the completion of their life cycle, although the earthworm is necessary for them to reach their final host; in such a situation the earthworm is termed a ***reservoir host.*** These parasites are taken into the body of the worm with food and soil; they accumulate in the earthworms body cavity, and often remain there for its whole life, which, for some species may be as long as several years (see Chapter 3). Thus, the earthworm enables the parasite to remain protected, in a situation in which it can retain its capacity for further development, and still be in a position to infect its final hosts. This is particularly true for those lumbricids which migrate from the surface to deeper soil in adverse conditions, and return to the surface soil in warm moist weather, when they are eaten by various species of birds, which then become infected with the parasites. A further point is that, while the parasite is in the body of the earthworm, it is protected against control measures used by man. Typical examples of parasites transmitted in this way are the nematodes (gape-worms) such as *Syngamus trachea* and *Cyathostoma bronchialis*, which parasitize the tracheae of various species of birds.

TABLE 24

Transmission of parasites by earthworms

Parasite	Earthworm	Final host	Reference
(i) *Protozoa*			
Histomonas sp. (blackhead)	—	Chickens (caecum)	Lund *et al* (1963)
(ii) *Cestoda* (tapeworms)			
Dilepus undula Schrank	*L. terrestris*	Birds and	
	"	rodents	Vogel (1921)
	"	"	Rysavy (1964)
Amoebotaenia cuneatus Linstow	*E. foetida*	Chickens	Grassi and Rovelli (1892)
	Pheretima sp.	"	Magalhaes (1892)
	P. peguana	"	Meggitt (1914)
	Ocnerodrilus africanus	"	Mönnig (1927)
	L. terrestris	"	Rysavy (1969)
A. lumbrici Villot	*Polycercus lumbrici*	Birds	Villot (1883)
Paricterotaenia paradoxa	*A. caliginosa*	Chickens	Genov (1963)
(iii) *Nematoda* (eelworms)			
Metastrongylus elongatus Duradin	*E. foetida*	Swine	Hobmaier and
M. pudentotectus Wostokow	*L. terrestris*	(lungs)	Hobmaier (1929)
M. salami Gedoelst (lungworms)	*L. rubellus*	"	Schwartz and
	A. caliginosa	"	Alicata (1931)
	E. veneta	"	Breza (1959)
Hystrichis tricolor Dujardin	*A. dubiosa*	Swine (stomach)	Rysavy (1969)
	Criodrilus lacuum	Ducks	Rysavy (1969)
		(oeso-phagus)	Karmanova (1959)
Capillaria annulata Molin	*E. foetida*	Chickens	
	A. caliginosa	(intestine)	Rysavy (1969)

TABLE 24—*continued*

Parasite	Earthworm	Final host	Reference
C. causinflata Molin	*A. caliginosa*	Chickens	Rysavy (1969)
	E. foetida	„	
	L. terrestris	„	
C. plica Rudolph	*L. terrestris*	Small predators	Rysavy (1969)
	L. rubellus	(urinary bladder)	
C. putorii Rudolph		„	Skarbilovic (1950)
C. mucronata	*L. rubellus*	„	„
Thominx aerophilus Creplin	*L. terrestris*	„	Rysavy (1969)
	L. rubellus	„	
	A. caliginosa	„	
Porrocaecum crassum Delongchamps	—	Birds	Mozgovoy (1952)
P. ensicaudatum Zeder	*L. herculeus*	„	Rysavy (1969)
Syngamus trachea Montagu *S. skrjabinomorpha* Ryzikov *S. merulae* Baylis *Cyathostoma bronchialis* Muhling	*E. foetida* *A. caliginosa* *A. longa* *L. terrestris*	Birds and chickens (trachea) „	Ryzhikov (1949) Ryzavy (1969)
Dioctophyma sp.	—	Mammals (kidneys)	Woodhead (1950) Karmanova (1963)
Spiroptera turdi	*L. terrestris*	Birds (stomach)	Cori (1898)
Stephanurus sp.	*E. foetida*	pig (kidney)	Tromba (1955)

Earthworms are also important as *passive agents* in the distribution of parasites. They may ingest the eggs of parasites with their food, carry them down to deeper soil, and protect them from harm caused by adverse physical factors. The eggs can pass through the intestines of earthworms without losing any of their infectivity. For instance, the eggs of *Ascaris suum* and *Ascaridia galli* passed through the intestine of individuals of *Lumbricus terrestris* without damage (Bejsovec, 1962). In this way, the eggs are spread throughout the soil, and facilitate infection of domestic animals and birds.

It has also been suggested that earthworms can infest human beings. *Microscolex modestus* has been reported from a fistula, *Eisenia foetida* in human urine, and *L. terrestris* and *Octolasium lacteum* from human faeces and the female vagina (Stolte, 1962).

10.5 Adverse effects on soil

Most reports in the literature state that earthworms improve soil structure, aeration and drainage, but a few workers have concluded that soil structure and plant growth can deteriorate as a result of the activities of earthworms. For instance, Agarwal *et al.* (1958) reported that a species of *Allolobophora*, by excreting a waxy fluid, made soil in parts of India cloddy and unproductive; this has not been substantiated and seems doubtful, because *Allolobophora* species are among the most common in fertile agricultural soils (Barley, 1959). Puttarudriah and Sastry (1961) claimed that the earthworms *Pontoscolex corethrurus* and *Pheretima elongata* had adverse effects on soil structure and the growth of some plants, in India. They stated that such large numbers of worms were present that the fine earth of their castings, which become mixed with intestinal secretions, mucus and other excreta had caused the soil to become compacted and cement-like. This resulted in the soil becoming very poorly drained and heavy, forming large clods. A number of test crops were grown, and although maize (*Zea mays*) and Ragi (*Eleusine coracana*) grew well, most vegetable crops such as carrots, radishes and beans did not do well, and had only very short tap roots.

Several workers have commented on the way earthworms contributed to soil erosion by bringing very finely divided soil to the surface. Darwin (1881) carefully observed the movement of earthworm casts down slopes, and Barley (1959) observed that this might eventually lead to soil erosion. Evans and Guild (1947) and Lee (1959) have reported that when there is excessive casting due to very large populations of earthworms, soil may be 'poached' (deep muddy impressions made by livestock.)

10.6 Earthworms as benefactors

The use of earthworms for soil improvement has been discussed in an earlier chapter, but earthworms also benefit man in other ways. They are widely used as bait for fish, and large commercial farms that produce earthworms for fish bait are found in the United States, as well as some in Great Britain. The commonest species used as bait is *L. terrestris*, but many other species, including particularly *Allolobophora caliginosa*, *Allolobophora chlorotica*, *E. foetida*, *Lumbricus rubellus*, *O. lacteum* and *Pheretima* spp., have also been used.

Earthworms have also been used for human food. They are regarded as a delicacy by the Maoris in New Zealand. In Japan, earthworm pies have been made, and there have been reports from South Africa of fried earthworms being eaten (Ljungström and Reinecke, 1969). Primitive natives from New Guinea and parts of Africa have been reported to eat raw earthworms.

There have been many reports of earthworms being eaten by humans as medicine, to cure such ills as stones in the bladder, jaundice, piles, fever and to alleviate impotency. Earthworm ashes have been used as a tooth powder in primitive societies (Stephenson, 1930), and it has been suggested that earthworms might contain a substance effective in curing rheumatism (Weisback, 1962).

Earthworms have been used in testing for pregnancy; urine from human females (concentrated according to Zonek 5 : 1) is injected into earthworms subcutaneously, and smears taken from their seminal vesicles, both before and after injection, to assess spermatogenesis. An accuracy of 90% was claimed for this method of pregnancy testing (Hasenbein, 1951).

A method of testing substances for carcinogenic properties was described by Gersch (1959) who found that benzopyrine (0·5%), dimethylbenzanthrene (0·5%) and other compounds, when applied to *L. terrestris* for several weeks, induced tumours.

11. Simple experiments and field studies with earthworms

11.1 Cultures

Earthworms can be kept in culture in almost any large vessel, provided that it is non-porous (otherwise it will be difficult to keep the soil sufficiently moist) and large enough, its actual size depending on the numbers of worms in the culture. Some species of earthworms produce their normal number of cocoons, only if given more space than they require to live and remain healthy. 'Stocks' of earthworms can be kept in large glazed earthenware pots, about 12 cm diameter and 30 cm deep, with the top covered with muslin, although for experimental purposes, smaller vessels are usually preferable, so that the amount of soil required is not excessive. Glass jars, such as 'Kilner' jars, having wide mouths are very suitable for earthworm experiments. Evans and Guild (1948) successfully maintained cultures for the study of cocoon production, with three earthworms per 5-pint (2·85-litre) capacity jar (*Lumbricus terrestris*), five per 3-pint (1·71-litre) capacity jar (*Allolobophora longa*) and five per 1-pint (450-ml) jar for smaller species such as *Allolobophora caliginosa*, *Allolobophora chlorotica* and *Allolobophora rosea*. Normally cultures should be kept in a cool building or cellar at a temperature between 10°C and 15°C or buried in soil. The ideal culture medium for most of the common pasture and garden species is a friable loam with 25–30% moisture content (such that it will remain in a ball when lightly squeezed in the hand). Food can be any partly-decomposed plant material, such as moist horse or bullock droppings (if these are dried before use, cocoons introduced with the food material will not be viable), partly rotted straw or tree leaves (not beech leaves). Cocoons can be obtained from field soil by washing it through a 1 or 2 mm sieve with a jet of water, so that they can then be picked off the sieve, or floated off by immersing the sieve in a solution of magnesium sulphate (specific gravity 1·2). Cocoons can also be kept in containers of soil, but if the time of hatching is to be studied, and to ensure the newly-emerged worms are not lost, they are better kept on moist filter paper in petri or other glass dishes, or under water in specimen tubes.

TABLE 25

The dimensions of the cocoons of sixteen species (mm)

	Mean Length (L)	Mean Width (W)	$\frac{L}{W}$
A. caliginosa	3·52	2·74	1·29
A. chlorotica	2·99	2·62	1·14
A. longa	6·20	4·34	1·43
A. nocturna	6·24	4·40	1·42
A. rosea	3·11	2·71	1·15
D. mammalis	2·24	1·96	1·14
D. rubida	2·11	1·75	1·20
D. subrubicunda	2·55	2·26	1·13
E. tetraedra	1·87	1·48	1·26
E. foetida	3·87	3·17	1·22
L. castaneus	2·35	2·06	1·14
L. festivus	3·52	2·88	1·22
L. rubellus	3·18	2·76	1·15
L. terrestris	5·97	4·69	1·28
O. lacteum	3·70	2·58	1·43
O. cyaneum	5·35	3·70	1·45

(*From Evans and Guild, 1947*)

Earthworms can be collected for culture purposes, either by digging for them or by using the formalin extraction method (see Chapter 5). Digging is simpler, and is satisfactory for obtaining small numbers of most surface-dwelling species, but it can be laborious, especially in pasture and more consolidated soils. If mature individuals of *L. terrestris* are required, the formalin method is more convenient. It may be necessary to cut long grass or rake away surface debris before applying the solution. It is most important, if the worms brought to the surface are not to be harmed by the formalin, that they should be washed immediately, or placed in a bowl of water before being put into soil in cultures.

11.2 Preservation

If the earthworms are for taxonomic study they can be killed by placing them either in a 5% formalin solution, or in fairly hot (50°C) water. They remain in a more relaxed condition if killed by the latter method, but formalin is more convenient and is also the most useful permanent preservative. Worms should be placed immediately in an appropriately-labelled jar containing formalin (1 lb screw-top honey jars are ideal for this purpose), so they can be left until it is convenient to examine them. Seventy per cent alcohol can also be used as a preservative, but the alcohol should be changed periodically, as the specimens dilute the solution with water from their bodies, and otherwise become macerated and unidentifiable.

11.3 Dissection

It is convenient, when pinning out earthworms for dissection, for study of the internal organs, or for identification, to insert the pins in the 1st, 5th, 10th and 15th segments, and so on. This facilitates the counting of the segments and recognition of the segmental position of the various organs.

11.4 Field experiments with earthworms

11.4.1 *Distributions*

Many field experiments on earthworms require only simple techniques. The basic technique recommended for estimation of populations is the formalin method, which may be combined with or replaced by the sampling and handsorting method described in Chapter 5. The geographical distribution of many species of lumbricids is very poorly known, and much valuable information can be obtained by sampling various habitats, such as pastures, arable land (under various crops), moorland, heathland, chalk downs, woodlands and various soil types. The best data concerning as many species as possible is probably obtained by combining the formalin- and handsorting methods. For each quadrat sampled with formalin, one or more soil samples of known volume are taken, and only *L. terrestris* individuals counted in the quadrat, whereas all the other species are counted in the soil samples. The two sets of data are then transformed into numbers per unit area and the results combined. However, this is rather tedious, and for general population studies, quadrats and formalin alone are usually sufficient. Before beginning a sampling programme, it is useful to take some trial samples and use both methods, comparing the results obtained with each.

11.4.2 *Effects of pH*

Soil pH is a factor that often greatly affects earthworm populations, both in numbers of individuals and numbers of species and this effect can be studied by comparing sampling results from different sites with soils of known pH. In general, there are fewer species in the more acid soils below pH 5 than in more alkaline soils, but further data is required.

11.4.3 *Production of casts*

Casts on the soil surface are a useful measure of the activity of the worms making them (particularly *Allolobophora nocturna* and *A. longa*). The number and weight of casts produced can be studied using quadrats similar in size, or larger, than those used for formalin. Several quadrats should be placed on the soil surface, and the casts occurring within each quadrat area removed daily, air or oven-dried and weighed. Alternatively, the number of times the worms cast during a longer period can be determined by starting with a cleared site, and observing the site every morning. The position of every *fresh* cast can then be plotted on to a sheet of squared paper, and the cast either removed, or marked with Indian ink, paint or dye, to avoid confusion with the next fresh cast. The two main casting species go into a diapause during the summer months, so it is an interesting exercise to study the effect of watering half the quadrats to see if this has any effect on the onset of diapause (represented by the time when casting ceases). The frequency and weight of casts can be compared in two or more different habitats, such as arable land and pasture.

11.4.4 *Incorporation of plant organic matter*

The activity of *L. terrestris* can be estimated from the amount of plant material gathered at the mouths of its burrows. An area should

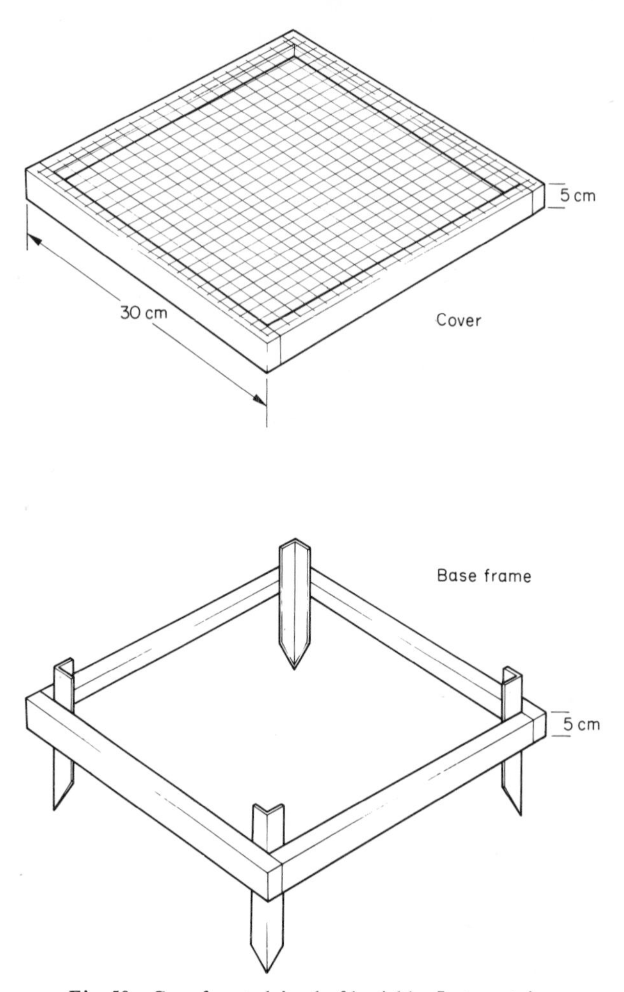

Fig. 59 Cage for studying leaf burial by *L. terrestris.*

be selected where the species is active, preferably one where leaves are obviously used as food material. Experimental cages are constructed, consisting of a wooden frame 30 cm square (i.e. the same size as a quadrat sample), joined at the corners by 5 × 5 cm angle-iron with points that can be driven into the ground (Fig. 59). This is placed on an area cleared of leaves, including any protruding from existing burrows. Leaves should be gathered from the ground at leaf-fall, and batches of one hundred leaves placed within each frame. A second wooden frame, covered with 2·5 cm mesh chicken wire is then placed on top of the first frame, to prevent the leaves being disturbed by birds or wind. This is made so the top frame can be removed without disturbing the position of the fixed frame. The total dry weight of the leaves laid out can be estimated by weighing sub-samples taken from the leaves left over from the original collection, and these are oven-dried at 80°C before reweighing. After one or two months, the leaves remaining can be removed, washed free of soil and also dried at 80°C and weighed. The weight of leaf material removed by the worms during this period can then be estimated and plotted on a graph. The amount of leaf material removed during weekly intervals, can also be estimated if sufficient leaves are collected at the beginning of the experiment to replenish the cages regularly and maintain the original number of leaves in each quadrat, including those leaves remaining and not pulled down. However, this does lead to biased results, as the amount of available food steadily decreases. At the end of the experimental period, the numbers of *L. terrestris* beneath the frames can be estimated by applying formalin, or by counting the burrow openings, or by both methods. The total number of leaves buried can be compared with the number of *L. terrestris* individuals beneath each cage, and also, if weekly observations are kept, correlated with climatic conditions such as temperature and rainfall.

11.4.5 *Palatability of leaves*

The palatability of different leaf material to earthworms, may be assessed by burying in soil, nylon bags of 7-mm mesh containing about fifty disks (2·5 cm diameter) cut, with a large cork borer, from fallen leaves. After one or two months, the leaves should be examined and the amount of leaf material taken by the worms assessed by washing the disks, then estimating the area of leaf remaining untouched when they are laid out on glass plates. This can be done by scoring a grid on a transparent plastic plate, and then holding each disk to a light with the grid behind or in front of it. With practice, reliable estimates of the amount of tissue that has disappeared can be made. After the estimation is complete, the disks are replaced in the mesh bags and buried again until the next sampling date. After several such estimates are made, a graph of the percentage breakdown against time can be plotted. An alternative method is to bury many more bags of disks, and at each sampling date dig up two or more of these, oven-dry the disks they contain at 80°C, and then weigh and discard them. Many sorts of leaf tissues can be used, such as kale, cabbage, beet and bean leaves, which disappear in one to two months, or maize, elm, ash, birch and lime, that are rather slower, and oak and beech which take one to two years to break down (Edwards and Heath, 1963).

An alternative experiment is to make up cultures with moist loam in buckets, then add ten large mature *L. terrestris* to each bucket. Place fifty leaves from each of any four different species of tree on the soil surface, and count the number of leaves remaining on the surface every three to seven days. From the results, percentage breakdown curves for the various species of leaves can be plotted.

11.4.6 *Burrowing experiments*

A useful observation cage for studying the habits of earthworms within the soil can be constructed from two sheets of glass slotted into a wooden framework so that they are about 1 cm apart (Fig. 60). The space between the sheets is filled with soil (sieved to remove large stones). After the frame is filled with soil, it should be tapped on the bench or floor to settle the soil. Earthworms can be introduced at the top, and allowed to burrow; they remain visible so their various activities can be studied as required, the cage being left covered with a dark cloth at other times. Studies can be made of the position in which cast material is deposited and the distribution of organic material through the soil. To study movement of organic matter through the soil, a layer of peat, farmyard manure, rotted straw, etc., should be placed on the top of the soil. The mixing of this layer with soil can be demonstrated by filling the rest of the frame with two or three layers of contrasting coloured soils, such as a clay subsoil, light sandy loam and a dark peaty loam. The open top of the frame can be covered with a strip of sticky transparent tape with holes punched in it to provide aeration. It is best to have different-sized frames for different species of earthworm. Small species, such as *A. chlorotica* and *A. rosea*, need only small frames

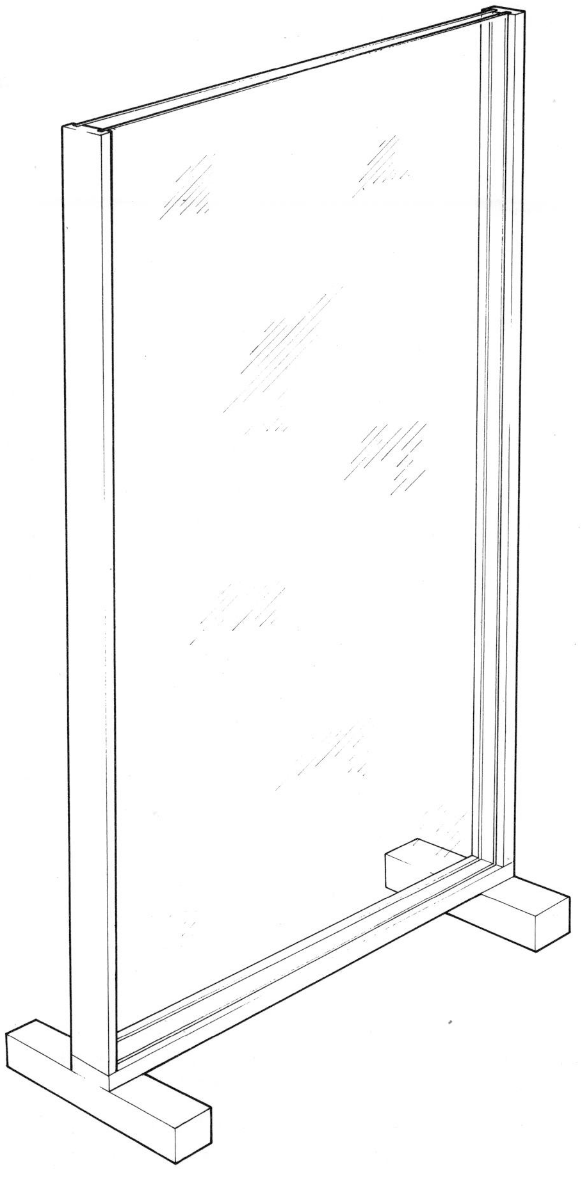

Fig. 60 Cage for studying earthworm activity.

about 30 cm deep by 22 cm wide, but *L. terrestris* needs larger frames up to 90 cm deep. A subsidiary study would be to follow the pattern of earthworm burrows in the field, by pouring latex down the exit holes and then digging up the solidified casts of the burrow system (Garner, 1953) (see Chapter 5).

11.4.7 *Life history studies*

The details of the life cycles of many species of earthworms are still obscure and many simple experiments can provide useful information.

A. Earthworms can be cultured in small containers of soil and provided with ample food, and their developmental cycles studied. For instance, if earthworms are kept from an early stage in small soil cultures outdoors and examined weekly, their growth, changes in segmentation, and sexual condition, can be recorded and also how soon they go into diapause noted. The results can be compared with observations on worms in cultures kept indoors or at any available

TABLE 26

Optimum temperatures for the development of earthworms

A. rosea	12°C
A. caliginosa	12°C
A. chlorotica	15°C
O. cyaneum	15°C
L. rubellus	15–18°C
D. attemsi	18–20°C
D. rubida	18–20°C
E. foetida	25°C

(*From Graff, 1953*)

controlled constant temperature (Table 26). Alternatively, development in cultures kept moist, and others that are gradually allowed to dry out, can be compared.

B. The numbers of cocoons produced when earthworms are fed on different diets can be studied in cultures (Fig. 61). A suitable number of mature (clitellate) earthworms, appropriate to the size of the culture vessel, are put in to cultures and fed on either farmyard manure, sheep dung or chopped straw, with at least two replicates of each treatment. After three months, cultures should be emptied and cocoons floated out of the soil as described earlier in the chapter. The number of eggs in some cocoons can be determined by dissec-

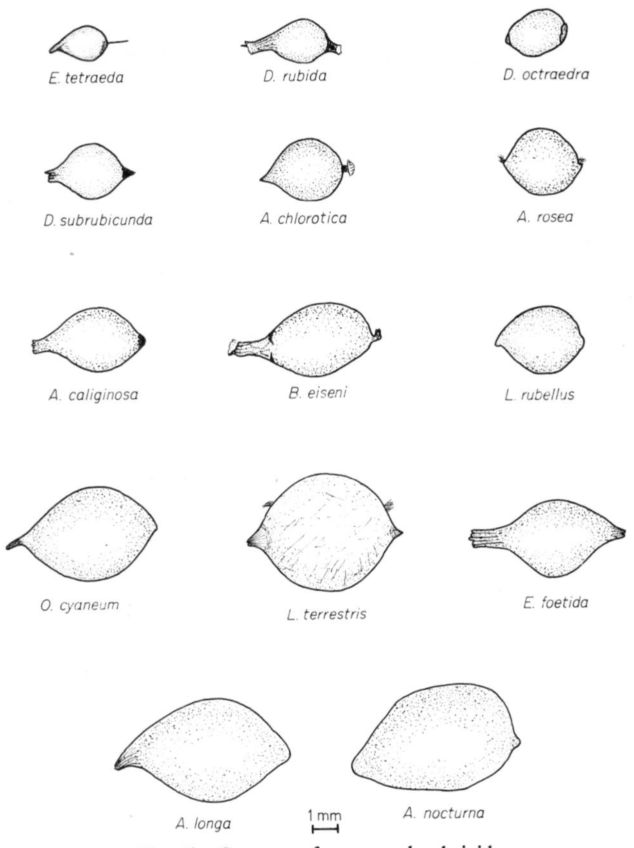

Fig. 61 Cocoons of common lumbricids. (*From various sources*)

tion, and others can be kept on moist filter paper and the numbers of worms hatching recorded. *A. chlorotica* and *Lumbricus castaneus* are particularly suitable for such studies, as they are small, do not go into obligatory diapause and reproduce rapidly. Nevertheless, information on other species is urgently needed.

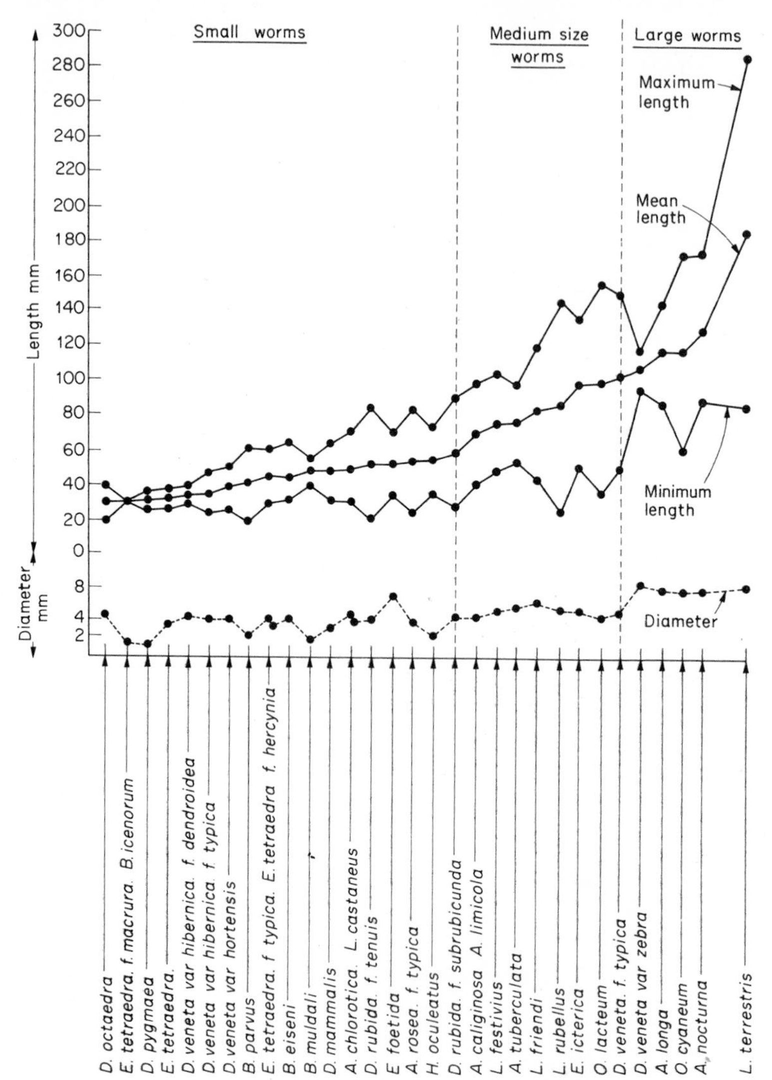

Fig. 62 The dimensions of British lumbricids.
(*Modified from Arthur, 1965*)

11.4.8 *Vertical and horizontal distribution of earthworms*

These can be studied by sampling a pattern of adjoining quadrats in 5 cm deep layers and handsorting the worms. After digging to a depth of 20–30 cm, the deep-burrowing species such as *L. terrestris*, can be brought up by pouring formalin solution on the soil after it has been levelled. They should be separated into immature, intermediate and sexually-mature individuals. Such studies demonstrate the aggregated patterns of distribution of earthworms.

11.4.9 *Effects of insecticides on earthworms*

The effects of insecticides can be studied in small plots of about 4 m^2. Any insecticide available can be tested either as a dust or liquid, and it is useful to compare its effect when thoroughly cultivated into the soil with those when it is left as a surface treatment. Usually, those species that come on to the soil surface at night are most affected by the surface treatments. Worms can be extracted from the plots at two or four-weekly intervals after treatment, using the formalin method, but the same quadrats should not be sampled twice.

11.4.10 *Physiological experiments*

Many simple physiological experiments are possible with earthworms. A few examples include:

A. Survival in water, with and without food, can be compared with survival in sugar solutions of different strengths, and survival in test chambers, in which different humidities are obtained by saturated salt solutions.

B. The reaction of worms to acid solutions can be studied by dipping the front end of worms into different acid solutions, and determining the minimum pH which will cause them to react.

C. The reaction of worms to moisture gradients in soil, ranging from dry to completely saturated soils produced by differential watering, can be recorded.

D. Different numbers of segments can be removed from either the anterior or posterior ends of worms and their regeneration followed step by step, either in soil cultures, which are examined every few days, or in dishes on moist blotting paper (food must be provided).

E. If worms are cut in two and their body walls joined by thread, tension on the worm causes contractions to pass along the body in waves. These experiments can be compared with the effect of cutting the body and leaving the nerve cord intact, removing the cerebral ganglia, and similar experimental treatments.

11.4.11 *Behavioural experiments*

A maze suitable for studying the learning ability of earthworms can be constructed, either from glass tubing or box-sections of sheet polythene, in the form of a T or Y. These shapes are designed to give the worm a choice of two pathways, towards or away from some stimulus, such as an electric shock produced by two electrodes. Although a box-sectioned maze may be easier to construct and manipulate, a tubular maze, which more nearly resembles the animal's natural burrows, is probably best, as the responses of the worm will be more normal (Swartz, 1929). The internal diameter or width can range from 0·5 to 2·0 cm, depending on the size of the worm. The maze is best constructed so that the arms can be easily separated from the 'stem', by making an oversize T or Y-shaped piece into which the arms and stem can be inserted (this will also have the advantage of making the junction larger than the maze components, thus giving the worm room to manoeuvre when turning into the side-arms). The floor of the maze is covered with damp filter paper to keep the atmosphere moist, and to allow the worm to move. The filter paper should be changed frequently to prevent the worms following previous tracks. The electrodes (or negative goal) are thin bare wires (0·5 cm apart), resting on a sheet of rubber introduced into one arm of the maze. The wires are parallel to the length of the arm, with their ends towards the junction, but sufficiently far away so that the worm does not reach them until it is well into the arm. The electrodes are connected to a 1·0–1·5 V battery, either permanently, so that the worm completes the electrical circuit on contact with them, or through a switch which can be manipulated when the worm is seen to touch the electrodes. Alternatively, an induction coil may be used instead to give more powerful, but intermittent, shocks. At the outer end of the other arm is placed a glass beaker or box, from which light has been excluded, and with its floor covered with damp moss and soil (the positive goal). The worm is introduced into the maze, and if it does not keep moving, it can be stimulated with a camel-hair brush. Once a worm has moved completely into an arm, it is removed, and the process repeated. A set of preliminary trials are made first without the electrodes, to see if the worm has any natural bias to turn left or right. Tests are then made with the electrodes inserted into the arm towards which the worm shows the most bias. After a number of repeated trials the worm should tend to turn more times into the other arm, as it begins to 'learn' a conditioned response to avoid entering the arm containing the electrodes. A large number of trials may be necessary before significant results are obtained. The 'learning' abilities of different species of earthworm can be compared, and possibly the effect of such factors as temperature, time of day, etc., on learning studied.

A simplified key to common genera of terrestrial earthworms

1. Clitellum commencing in front of the 15th segment (not found in the British Isles)	2
— Clitellum commencing after the 15th segment	3 (LUMBRICIDAE)
2. Setal arrangement perichaetine (Fig. 2)	*Pheretima* (MEGASCOLECIDAE)
— Setal arrangement lumbricine (Fig. 2)	*Diplocardia* (ACANTHODRILIDAE)
3. Prostomium tanylobous (fig. 1), setae closely paired (Fig. 2) at least over part of the body	4
— Prostomium epilobous (fig. 1) or if tanylobous, setae widely paired or distant (Fig. 2) over the whole body	5
4. Clitellum begins on segment 24	*Bimastos eiseni* (Levinsen)**
— Clitellum begins on or after segment 26	*Lumbricus*
5. Clitellum ends after segment 28	6
— Clitellum ends before segment 28	*Eiseniella*
6. Tubercula pubertatis absent, or exceptionally, present as simple thickenings of the edges of the clitellum	*Bimastos*
— Tubercula pubertatis present as ridges or isolated papillae (plate 5b)	7
7. Setae widely-paired or distant, at least posteriorly	8
— Setae closely-paired throughout the length of the body	9
8. Setae widely paired or distant throughout the length of the body, tubercula pubertatis as ridges (exceptionally as separate tubercles) extending over only part of the length of the clitellum	*Dendrobaena*

** Only one common species.

— Setae closely-paired anteriorly (in the region of the hearts), distant posteriorly. Tubercula pubertatis as ridges as long or longer than the clitellum — *Octolasium*

9. Spermathecal pores (Fig. 3) in line with seta ‘d’ or more often near the mid-dorsal line, body trapezoidal in crossection — *Eisenia*

— Spermathecal pores situated laterally between setae ‘c’ and ‘d’ or ‘a’ and ‘b’ and ‘c’ and ‘d’. Body not trapezoidal in crossection — 10

10. Prostomium with longitudinal ridges — *Eophila*

— Prostomium without longitudinal ridges — 11

11. Calciferous glands with two lateral pouches in segment Terrestrial — *Allolobophora*

— Calciferous glands without lateral pouches. Amphibious — *Helodrilus*

Note. Internal characteristics have been used only in couplet 11. However, most *Helodrilus* species are amphibious, living in mud at the bottom of ditches, ponds, etc., and in wet river banks, whereas *Allolobophora* species are terrestrial. Therefore the habitat can usually be used as the key character thus avoiding the necessity of dissection.

Simple key to species of terrestrial earthworms

Most common and widespread species of earthworms can be identified using the characters in the first two or three columns of the key. If rarer species that are not included in the key are found, they may be misidentified on the basis of the characters in the first three columns, but reference to the additional descriptions given should usually be sufficient to avoid confusion.

MEGASCOLECIDAE

Genus *Pheretima*

Two pairs of spermathecal pores in segmental grooves 7/8 and 8/9. 1st dorsal pore 11/12. 70–170 mm. 10–150 segs. Reddish-brown, clitellum creamy to dark grey. — *Pheretima californica* (Kinberg)

Three pairs of spermathecal pores on the anterior edges of segments 7, 8 and 9. 150–220 mm. Light green/greenish buff with purple green dorsal line, clitellum milky or chocolate, pale grey ventrally. — *P. hupiensis* (Michaelsen)

Four pairs of spermathecal pores in segmental grooves 5/6, 6/7, 7/8 and 8/9. 1st dorsal pore 12/13. 100–160 mm. 120–150 segs. Rich brown. — *P. posthuma* (Vaillant)

ACANTHODRILIDAE

Genus *Diplocardia*

1. Clitellum forms a complete ring around the body. 40–120 mm. 90–120 segs. Anterior dorsal surface pale flesh coloured. — *Diplocardia singularis* (Ude)

Clitellum not a complete ring, but saddle-shaped.

2. Three pairs of spermathecal pores in segmental grooves 6/7, 7/8 and 8/9. 180–300 mm. 125–160 segs. Anterior dorsal surface pale flesh coloured. *D. communis* (Garman)

Two pairs of spermathecal pores in segmental grooves 7/8 and 8/9. 200–270 mm. 135–160 segs. Anterior dorsal surface dark brown. *D. riparia* (Smith)

LUMBRICIDAE

Genus *Lumbricus*

Clitellum†	*Tubercula*† *pubertatis*	*1st dorsal*† *pore*		
26, 27–32	28–31	7/8	Red/brown or red/violet, irridescent dorsally, pale yellow ventrally. 25–105 mm. 95–120 segs.	*Lumbricus rubellus** Hoffmeister
28–33	29–33	6/7	Chestnut to violet brown; brown/yellow ventrally, irridescent, clitellum orange. 30–70 mm. 82–100 segs.	*L. castaneus** (Savigny)
31, 32–37	33–36	7/8	Setae widely-paired both ends of the body, strongly pigmented, brown-red dorsally, yellowish ventrally. 90–300 mm. 110–160 segs.	*L. terrestris** Linnaeus
33, 34–39	34, 35, 36–38	5/6	Red-brown, lighter ventrally, irridescent dorsally, prominent clitellum. Not found in large numbers. 48–108 mm. 100–143 segs.	*L. festivus** (Savigny)

Genus *Eiseniella*

22, 23–26, 27	23–25, 26	4/5	Male pores in 13. Dark brown, greenish, golden yellow, red. Body quadrangular behind the clitellum. 30–60 mm. 60–90 segs.	*Eiseniella tetraeda f. typica** (Savigny)
,,	,,	,,	Male pore in 15. Otherwise as f. *typica*	*E. tetraeda* (Savigny) *f. hercynia** (Michaelsen)

* British species. † for numbering see p. 12.

Genus *Bimastos*

Clitellum	*Tubercula pubertatis*	*1st dorsal pore*		
20, 21, 22–29, 30	Absent		Setae ab > cd. Red-brown. 105–115 segs. 50–80 mm.	*Bimastos gieseleri* (Smith)
22–29	27, 28		Reddish-brown. Eise segs 30–6. 20–50 mm	*B. tumidis*
23–28	Absent		20–50 mm. 40–60 segs. up to 75 mm.	*B. palustris* (Moore)
23, 24, 25–31, 32	24, 25, 26–30 or absent	5/6	Reddish dorsally, yellowish ventrally. 25–40 mm. 90–110 segs.	*B. parvus* (Eisen)
23–32 or 24–33	Absent		Rose-red. 98–122 segs. 60–90 mm.	*B. longicinctus* (Smith & Gittins)
24, 25–32, 33	Absent	5/6	Prostomium tanylobous, body cylindrical, reddish or violet dorsally, yellowish ventrally, clitellum red. 30–64 mm. 75–111 segs.	*B. eiseni* * (Levinsen)
25, 26, 27–30, 31, 32, 33	28, 29–30, 31 or absent	5/6	Setae widely paired, dorsally red-brown with bluish tint. Intersegmental grooves and ventral side light. 20–85 mm. 90–120 segs.	*B. tenuis* *† (Eisen)
27–37	Absent		Pale red to chestnut brown, often localized whitish banding anteriorly. 100–140 mm. 110–140 segs.	*B. zeteki* (Smith & Gittins)

Genus *Dendrobaena*

Clitellum	*Tubercula pubertatis*	*1st dorsal pore*		
25, 26–31, 32	28–30	5/6	Rosy to deep red, last posterior segments yellow. 27–90 mm. 50–100 segs.	*Dendrobaena subrubicunda*‡ (Eisen)

* British species. † Now *Dendrobaena rubida* (Sav.) f. *tenuis* (Eisen).
‡ Now *Dendrobaena rubida* (Sav.) f. *subrubicunda* (Eisen).

Clitellum	*Tubercula pubertatis*	*1st dorsal pore*		
25, 26–28, 29, 30	25–30	5/6 or sometimes 8/9	Smoky-grey with red pigment posteriorly. 80–180 mm. 120–160 segs.	*D. platyura* (Fitzinger)
26, 27–31, 32	29–30	5/6	Dark red dorsally, lighter red ventrally. 30–60 mm. 50–100 segs.	*D. rubida** (Savigny)
24, 25–26, 27–32, 33	30 and 31	5/6	Violet, purple or olive brown, dorsal pigment bands separated by non-pigmented zones. Sometimes uniformly unpigmented. 50–155 mm. 80–225 segs.	*D. veneta f. typica* (Rosa)
27, 28, 29–33, 34	31–32, 33	4/5	Red, violet, yellow or copper. Posterior octagonal. 17–40 mm. 79–95 segs.	*D. octaedra** (Savigny)
28–33, 34	29, 30–32	5/6	Dorsally pale red, 1st segment and ventrally and clitellum white. 20–50 mm. 100–150 segs.	*D. attemsi* (Michaelsen)
31–36	33–34	4/5	Red-violet, slightly irridescent. 30–65 mm. 83–100 segs.	*D. mammalis** (Savigny)
33–37	34, 35–36, 37	Undetectable	Red dorsally or unpigmented. 30–32 mm. 103–180 segs.	*D. pygmaea** (Savigny)
Genus *Octolasium*				
29–34	30–33	11/12	Blue-grey with (usually) lilac-blue dorsal line. Last 4–5 segments yellow, anterior segments pink, clitellum red-orange. 50–160 mm. 100–150 segs.	*Octolasium cyaneum* (Savigny)
30–35	31–34	8/9, 9/10 or 10/11	White, grey, blue or rose-pink, clitellum pink or orange. 30–160 mm. 90–180 segs.	*O. lacteum** (Oerley)

* British species.

Clitellum	*Tubercula pubertatis*	*1st dorsal pore*		
Genus *Eisenia*				
24, 25, 26–32	28–30, 31	4/5	Red, purple or brown; yellowish ventrally. Dorsal surface pigment alternating with light intersegmental zones. 32–130 mm. 80–110 segs.	*Eisenia foetida** (Savigny)
Genus *Eophila*				
33, 34 35–42, 43, 44	35, 36, 37–41, 42, 43, 44		Yellowish or grey. 52–140 mm. 132–170 segs.	*Eophila icterica** (Savigny)
Genus *Allolobophora*				
25, 26–31, 32, 33	29–30, 31 or 30–32	4/5	Pale red, without pigment, prominent and flattened clitellum. 25–85 mm. 120–150 segs.	*Allolobophora rosea** (Savigny)
26–32	28, ½29 and ½29, 30	5/6	Grey, unpigmented, usually blood shows through epidermis. 20–60 mm. 80–120 segs.	*A. culpifera* (Tétry)
25, 26–33	30 and 31	4/5	Whitish grey, unpigmented. 50–90 mm. 100–130 segs.	*A. antipae* (Michaelsen)
26, 27, 28–34, 35	31 and 33	11/12 or 12/13	Anterior, especially the first few segments pink, otherwise pale grey, yellowish clitellum. 40–100 mm. 120–150 segs.	*A. caliginosa f. typica** (Savigny)
26, 27 28–34, 35	31–33		Colour as for f. *typica*. Tubercula pubertatis of two raised tubercles, connected by a narrow bridge.	*A. caliginosa* (Savigny) *f. trapezoides* (Duges)
27–32, 33	Absent	Indistinct	Grey, unpigmented. 80–100 segs. 22–25 mm.	*A. miniscula* (Rosa)
27, 28–35	32 and 34		Body cylindrical, pale grey, unpigmented. 90–150 mm. 160–200 segs.	*A. terrestris* (Savigny)

* British species.

Clitellum	Tubercula pubertatis	1st dorsal pore		
27, 28–35	32–34		Body cylindrical, colour as for *A. terrestris*. 90–150 mm. 171–181 segs.	*A. longa** Ude
27, 28–35	31 and 33 (extending into segment 32)		Segments posterior to 13 divided by two grooves into three rings. Dark reddish brown, clitellum paler. 90–180 mm. 200–250 segs.	*A. nocturna** Evans
27, 28–34, 35	31–35	10/11, 11/12 or 12/13	55–100 mm. 152–194 segs. Body cylindrical, unpigmented, greyish.	*A. tuberculata* Eisen
28–35, 36	33 and 34	4/5 or 5/6	Unpigmented, anterior pink, the rest of the body pinkish grey. Bulbous anterior. 40–100 mm. 86–146 segs.	*A. limicola** Michaelsen
28, 29–37	31 and 33 and 35	4/5	Light or dark green, yellow, grey, pink, slate-blue, clitellum pink, green or grey. 30–70 mm. 80–138 segs.	*A. chlorotica** (Savigny)
Genus *Helodrilus*				
21, 22–32	29–30	4/5	Setae black in fully mature individuals, flesh-coloured, body unpigmented. 35–75 mm. 95–150 segs.	*Helodrilus oculatus** Hoffmeister

* British species.

References

Agarwal, G. W., Rao, K. S. K. and Negi, L. S. (1958). Influence of certain species of earthworms on the structure of some hill soils. *Curr. Sci.* **27,** 213.

Aichberger, R. von (1914). Untersuchungen über die Ernährung des Regenwurmes. *Ztsch. Deutsch. Mikrob. Gesell.* **58,** 69–72.

Aichberger, R. von (1914). Studies on the nutrition of earthworms. *Kleinwelt,* **6,** 53–8, 69–72, 85–8.

Aisyazhnyuk, A. A. (1950). Use of 666 for the control of chafer grubs. *Agrobiologiya,* **5,** 141–2.

Allee, W. C., Torvik, M. M., Lahr, J. P. and Hollister, P. L. (1930). Influence of soil reaction on earthworms. *Physiol Zool.* **3** (2) 164–200.

Allen, R. W. (1960). Relative susceptibility of various species of earthworms to the larvae of *Capillaria annulata. Proc. Helminthol. Soc. Wash.* **17** (2) 58–64.

Anstett, M. (1951). Sur l'activation macrobiologique des phénomènes d'humification. *C.R. Hebd. Seanc. Acad. Agric. France,* 230.

Arbit, J. (1957). Diurnal cycles and learning in earthworms. *Am. Assoc. Adv. Sci.* **126,** 654–5.

Arldt, T. (1908). Die Ausbreitung der terricolen Oligochaeten im Laufe der erdgeschichtlichen Entwicklung des Erdreliefs. *Zool. Jahrb. Syst.* **26.**

Arldt, T. (1919). *Handbuch der Palaeogeographie.* Leipzig.

Arrhenius, O. (1921). Influence of soil reaction on earthworms. *Ecology,* **2,** 255–7.

Arthur, D. R. (1965). Form and function in the interpretation of feeding in Lumbricid worms. *Viewpoints in Biology,* **4,** 204–51.

Atlavinyte, O. (1964). Distribution of earthworms (Lumbricidae) and larvae of insects in the eroded soil under cultivated crops. *Pedobiologia,* **4,** 245–50.

Atlavinyte, O. (1965). The effect of erosion on the population of earthworms in the soils under different crops. *Pedobiologia,* **5,** 178–88.

Atlavinyte, O. and Lugauskas, A. (1972). The effect of Lumbricidae on the soil microorganisms. *Proc. IVth Int. Symp. Soil Zool.* (in press).

Avel, M. (1959). Classe des Annélides Oligochaetes (Oligochaeta. Huxley, 1875). *Traite de Zoologie*, **5** (1), 224–271.

Bachelier, G., (1963). *La Vie Animale dans les Sols.* Orstom, Paris. 279 pp.

Bahl, K. N. (1919). On a new type of nephridia found in Indian earthworms of the genus *Pheretima. Q. Jl. micros. Sci.* **64**.

Bahl, K. N. (1922). On the development of the 'enteronephric' type of nephridial system found in Indian earthworms of the genus *Pheretima. Q. Jl. Micros. Sci.* **66,** 49–103.

Bahl, K. N. (1927). On the reproductive processes of earthworms: Pt I. The process of copulation and exchange of sperm in *Eutyphoeus waltoni. Q. Jl. micros. Sci.* **71,** 479–502.

Bahl, K. N. (1947). Excretion in the Oligochaeta. *Biol Rev.* **22,** 109–47.

Bahl, K. N. (1950). *The Indian Zoological Memoirs. I. Pheretima.* 4th edition. Lucknow Pub. House, Lucknow.

Baker, W. L. (1946). D.D.T. and earthworm populations. *J. econ. Ent.* **39,** 404–5.

Bakhtin, P. U. and Polsky, M. N. (1950). The role of earthworms in structure formations of sod-podzolized soils. *Pochvovedenie,* 487–91.

Baldwin, F. M. (1917). Diurnal activity of the earthworm. *J. Anim Behav.* **7,** 187–90.

Ball, R. C., and Curry, L. L. (1956). Culture and agricultural importance of earthworms. *Mich. Stat. Univ. Agr. Exp. Stn. Coop. Ext. Soc. Circ. Bull.* **222.**

Baluev, V. K. (1950). Earthworms of the basic soil types of the Iranov region. *Pochvovedenie,* 487–91.

Barker, R. J. (1958). Notes on some ecological effects of DDT sprayed on elms. *J. Wildl. Manage.* **22,** 269–74.

Barley, K. P. (1959). The influence of earthworms on soil fertility. II. Consumption of soil and organic matter by the earthworm *Allolobophora caliginosa. Aust. J. agr. Res.* **10** (2) 179–158.

Barley, K. P. (1959). Earthworms and soil fertility. IV. The influence of earthworms on the physical properties of a red-brown earth. *Aust. J. agr. Res.* **10** (3) 371–6.

Barley, K. P. (1961). The abundance of earthworms in agricultural land and their possible significance in agriculture. *Adv. Agron.* **13,** 249–68.

Barley, K. P. and Jennings, A. C. (1959). Earthworms and soil fertility. III. The influence of earthworms on the availability of nitrogen. *Aust. J. agr. Res.* **10** (3), 364–70.

Barley, K. P. and Kleinig, C. R. (1964). The occupation of newly irrigated lands by earthworms. *Aust. J. Sci.* **26** (9) 290.

Barrett, T. J. (1959). Harnessing the Earthworm. Faber and Faber, London. 166 pp.

Bassalik, K. (1913). On silicate decomposition by soil bacteria. *Z. Gärungs-phyiol.* **2,** 1–32.

Bather, E. A. (1920). *Pontoscolex latus*, a new worm from Lower Ludlow, Beds. *Ann. Mag. Nat. Hist.* (9) 5.

Bauer, K. (1964). Studien über Nebenwirkungen von Pflanzenschutzmitteln auf die Bodenfauna. *Mitt. Biol. Bund. Land. Forst. Berlin-Dahlem.* **112,** 42 pp.

Buahin, G. K. A. and Edwards, C. A. (1964). The recolonisation of sterilised soil by invertebrates. *Rep. Rothamsted exp. Stn. for 1963,* 149–50.

Baweja, K. D. (1939). Studies of the soil fauna with special reference to the recolonisation of sterilised soil. *J. Anim. Ecol.* **8** (1), 120–61.

Baylis, H. A. (1914). Preliminary account of *Aspidodrilus*, a remarkable epizoic oligochaete. *Ann. Mag. Nat. Hist.* (8) 16.

Baylis, H. A. (1915). A new African earthworm collected by Dr C. Christy. *Ann. Mag. Nat. Hist.* (8) 16.

Beauge, A. (1912). Les vers de terre et la fertilité du sol. *J. Agric. prat. Paris.* **23,** 506–7.

Bejsovec, J. (1962). Rozsirovani Zarodu Helmintu Pasazi Zazivacim Traktem Adekvatnich Prenasecu. *Cs. Parasitol.* **9,** 95–109.

Benham, W. B. (1896). On *Kynotus cingulatus*, a new species of earthworm from Imerina in Madagascar. *Q. Jl. Micros. Sci.* p. 38.

Benham, W. B. (1922). Oligochaeta of Macquarie Island. *Australian Antarctic Expedition. Sci. Reports, Zool. and Bot.* **6.**

Bigger, J. H. and Decker, G. C. (1966). Controlling root-feeding insects on corn. *Illinois Univ. Agr. Exp. Stn. Bull.* **716,** 24 pp.

Bharucha-Reid, R. P. (1956). Latent learning in earthworms. *Science,* **123,** 222.

Blancke, E. and Giesecke, F. (1923). Mono- und Dimethyloharnstoffe in Ihrer Wirkung auf die Pflanzenproduktion und ihr Umsatz im Boden. *Z. Pflanz. Düng., Bodenkunde,* **2.**

Blancke, E. and Giesecke, F. (1924). The effect of earthworms on the physical and biological properties of soil. *Z. Pflanz. Düng., Bodenkunde,* **3** (B) 198–210.

Blankwaardt, H. F. H. and van der Drift, J. (1961). Invloed van Grondontsmetting in Kassen op Regenwormen. *Meded. Dir. Tuinbouw.* **24,** 490–6.

Block, W. and Banage, W. B. (1968). Population density and biomass of earthworms in some Uganda soils. *Rev. Ecol. Biol. Sol.* **5** (3) 515–21.

Bocock, K. L., Gilbert, O., Capstick, C. K., Twinn, D. C., Waid, J. S. and Woodman, M. G. (1960). Changes in leaf litter when placed on the surface of soils with contrasting humus types. I. Losses in dry weight of oak and ash leaf litter. *Soil Sci.* **11,** 1–9.

Bodenheimer, F. S. (1935). Soil conditions which limit earthworm distribution. *Zoogeographica,* **2,** 572–8.

Bornebusch, C. H. (1930). The fauna of the forest soil. *Forstl. Forsøgsv. Dan.* **11,** 1–224.

Bornebusch, C. H. (1953). Laboratory experiments on the biology of worms. *Dansk Skovforen. Tidsskr.* **38,** 557–79.

Bouché, M. B. (1966). Sur un nouveau procédé d'obtention de la vacuité artificielle du tube digestif des lumbricides. *Rev. Ecol. Biol. Sol.* **3** (3) 479–82.

Bouché, M. B. (1969). Comparison critique de methodes d'evaluation des populations de lumbricides. *Pedobiologia,* **9,** 26–34.

Boyd, J. M. (1957). The Lumbricidae of a dune-machair soil gradient in Tiree, Argyll. *Ann. Mag. Nat. Hist.* **12** (10) 274–82.

Boyd, J. M. (1957a). The ecological distribution of the Lumbricidae in the Hebrides. *Proc. R. Soc. Edinb.* **66,** 311–38.

Boyd, J. M. (1958). The ecology of earthworms in cattle-grazed machair in Tiree Argyll. *J. Anim. Ecol.,* **27,** 147–57.

Boynton, D. and Compton, O. C. (1944). Normal seasonal changes of oxygen and carbon dioxide percentages in gas from the larger pores of three orchard subsoils. *Soil Sci.* **57,** 107–17.

Bray, J. R. and Gorham, E. (1964). Litter production in forests of the world. *Adv. Ecol. Res.* **2,** 101–57.

Bretnall, G. H. (1927). Earthworms and Spectral Colours. *Science,* **66,** 427.

Bretscher, K. (1896). The Oligochaeta of Zürich. *Rev. Suisse Zool.* **3,** 499–532.

Breza, M. (1959). Kebologichym viztahom daziloviek (Lumbricidae) abo medzihostitelov preuno helmintov z rodu *Metastrongylus.* I. Novy unimavy druh medzihostitelov. *Eisenia veneta* (Rora) var. *hortensis* (Mich.). *Fol. veter. cas.* **3,** 251–66.

Brown, B. R., Love, C. W. and Handley, W. R. C. (1963). Protein-fixing constituents of plants: *Rep. For. Res. London, Part III,* 90–3.

Brown, D. M. (1944). The cause of death in submerged worms. *J. Tenn. Acad. Sci.* **19** (2) 147–9.

Bruel, W. E. van der (1964). Le sol, la pedofauna et les applications de pesticides. *Annales de Gembloux,* **70,** 81–101.

Brüsewitz, G. (1959). Untersuchungen über den Einfluss des Regenwurms auf Zahl und Leistungen von Mikroorganismen im Boden. *Arch. Microbial,* **33,** 52–82.

Buntley, C. J. and Papedick, R. I. (1960). Worm-worked soils of Eastern South Dakota, their morphology and classification. *Soil Sci. Soc. Amer. Proc.* **24,** 128–32.

Byzova, Yu B. (1965). Comparative rate of respiration in some earthworms. *Rev. Ecol. Biol. Sol.* **2,** 207–16.

Carter, G. S. (1940). *A General Zoology of the Invertebrates.* 4th edition. 421 pp.

Cavsey, D. (1961). The earthworms of Arkansas, in *The challenge of earthworm research,* Ed. R. Rodale. Soil and Health Foundation, Penn., pp. 43–52.

Cernosvitov, L. (1928). Eine neue, an Regenwürmern schmarotzende Enchytraidenart. *Zool. Anz.* **78.**

Cernosvitov, L. (1930). Oligochaeten aus Turkestan. *Zool. Anz.* **91** (1–4) 7–15.

Cernosvitov, L. (1930). Prispevky k poznani fauny tatranskych Oligochaetu. *Vestniku Kral Ces. Spol. Nauk,* **2,** 1–8.

Cernosvitov, L. (1931). Revision des *Lumbricus submontanus* Vejdovsky, 1875. *Zool. Anz.* **95** (1–2) 59–62.

Cernosvitov, L. (1931). Zur Kenntnis der Oligochaeten fauna des Balkans. *Zool. Anz.* **95,** (11–12) 312–27.

Cernosvitov, L. (1931). Eine neue *Lumbricus.* Art aus der Umgebung von Prag. *Zool. Anz.* **96** (7–8) 201–4.

Cernosvitov, L. and Evans, A. C. (1947). *Synopses of the British Fauna* (6) *Lumbricidae. Linn. Soc. London.*

Chadwick, L. C. and Bradley, J. (1948). An experimental study of the effects of earthworms on crop production. *Proc. Amer. Soc. hort. Sci.* **51,** 552–62.

Chapman, G. (1950). On the movement of worms. *J. exp. Biol. Cambridge,* **27,** 29–39.

Chen, C. M. and Liv, C. L. (1963). Dynamics of the populations and communities of rice insect pests in the bank of Fung-Ting Lake region, Hunar. *Acta. ent. Sin.* **12,** 649–57.

Cockerell, T. D. A. (1924). Earthworms and the cluster fly. *Nature, Lond.* **113** (2832) 193–4.

Cohen, S. and Lewis, H. B. (1949). Nitrogenous metabolism of the earthworm. (*L. terrestris*). *Fedn. Proc. Fedn. Am. Soc. exp. Biol.* **8,** 191.

Coin, C. J. (1898). Beitrag zur Biologie von *Spiroptera turdi. Sitzsser. Deutsch. nat.-med. Ver. Bohmen, Prag.*

Combault, A. (1909). Contribution a l'étude de la respiration et la circulation des Lombriciens. *J. Anat. Paris,* **45.**

Cohen, S. and Lewis, H. B. (1949). The nitrogen metabolism of the earthworm. *J. biol. Chem.* **180,** 79–92.

Cragg, J. B. (1961). Some aspects of the ecology of moorland animals. *J. Anim. Ecol.* **30,** 205–54.

Cramp, S., Conder, P. J. and Ash, J. S. (1965). *5th Rep. Joint Comm. of Brit. Trust. Ornith. and R.S.B.P. on Toxic Chemicals*, 20.

Crompton, E. (1953). Grow the soil to grow the grass. Some pedological aspects of marginal land improvement. *J. Minist. Agric. Fish.* **50** (7) 301–8.

Crossley, D. A., Reichle, D. E. and Edwards, C. A. (1971). Intake and turnover of radioactive cesium by earthworms (Lumbricidae). *Pedobiologia,* **11,** 71–6.

Darwin, C. (1881). *The formation of vegetable mould through the action of worms, with observations of their habits.* Murray, London. 326 pp.

Davey, S. P. (1963). Effects of chemicals on earthworms: a review of the literature. Special Scientific Report. Wildlife 74. U.S.D.I. Fish and Wildlife Service.

Davis, B. N. K. (1968). The soil macrofauna and organochlorine residues at twelve agricultural sites near Huntingdon. *Ann. appl. Biol.* **61,** 29–45.

Davis, B. N. K. and French, M. C. (1969). The accumulation and loss of organochlorine insecticide residues by beetles, worms and slugs in sprayed fields. *Soil Biol. Biochem.* **1,** 45–55.

Davis, B. N. K. and Harrison, R. B. (1966). Organochlorine insecticide residues in soil invertebrates. *Nature, Lond.* **211,** 1424–5.

Dawson, A. B. (1920). The intermuscular nerve cells of the earthworm. *J. Comp. Neurol.* **32,** 155–71.

Dawson, R. C. (1947). Earthworm microbiology and the formation of water-stable aggregates. *Soil Sci.* **69,** 175–84.

Dawson, R. C. (1948). Earthworm microbiology and the formation of water-stable soil aggregates. *Proc. Soil. Sci. Soc. Am.* **12,** 512–16.

Day, G. M. (1950). The influence of earthworms on soil micro-organisms. *Soil Sci.* **69,** 175–84.

Devigne, J. and Jevniaux, C. (1961). Sur l'origine des chitinases intestinales des lombrics. *Arch. int. Physiol. Biochim.* **68** (5) 833–4.

Dhawan, C. L., Sharma, R. L., Singh, A. and Handa, B. K. (1955). Preliminary investigations on the reclamation of saline soils by earthworms. *Proc. natn. Inst. Sci. India,* **24,** 631–6.

Dhennin, L. *et al.* (1963). Investigations on the role of *Lumbricus terrestris* in the experimental transmission of foot and mouth disease virus. *Bull. Acad. Vet. France,* **36,** 153–5.

Doane, C. C. (1962). Effects of certain insecticides on earthworms. *J. econ. Ent.* **55,** 416–18.

Dobson, R. M. (1956). *Eophila oculata* at Verulamium: a Roman earthworm population. *Nature, Lond.* **177,** 796–7.

Dobson, R. M. and Lofty, J. R. (1956). Rehabilitation of marginal grassland. *Rep. Rothamsted exp. Stn. for 1955.*

Dobson, R. M. and Lofty, J. R. (1965). Observations of the effect of BHC on the soil fauna of arable land. *Congr. Int. Sci. Sol. Paris,* **3,** 203–5.

Doeksen, J. (1950). An electrical method of sampling soil for earthworms. *Trans. 4th Int. Congr. Soil Sci.* 129–31.

Doeksen, J. and Couperus, H. (1962). An estimation of the growth of earthworms. *Wageningen Inst. v. Viol. en Scheik. Onderz. van Landgervassen. Meded.* **195,** 173–5.

Doeksen, J. (1964). Notes on the activity of earthworms. 1. The influence of *Rhododendron* and *Pinus* on earthworms. *Jaarb. I.B.S.* 177–80.

Doeksen, J. (1964). Notes on the activity of earthworms. 3. The conditioning effect of earthworms on the surrounding soil. *Jaarb. I.B.S.* 187–91.

Doeksen, J. (1967). Notes on the activity of earthworms. V. Some causes of mass migration. *Meded. Inst. biol. Scheik. Ouderz LandbGewass.* **353,** 199–221.

Doeksen, J. (1968). Notes on the activity of earthworms. VI. Periodicity in the oxygen consumption and the uptake of feed. *Meded. Inst. biol. Scheik. Onderz. LandbGewass,* **354,** 123–8.

Doeksen, J. and Couperus, H. (1968). Met vastellen van groei bij regerwormen. *Jaarb. I.B.S.* 173–5.

Doeksen, J. and van der Drift, J. (1963). Proceedings of the Colloquium on Soil Fauna Soil Microflora and their relationships. *Soil Organisms.* Oosterbeek, North Holland Pub. Co., Amsterdam, The Netherlands. 453 pp.

Doeksen, J. and Minderman, G. Typical soil structures as the result of the activities of mudworms. (mimeo publication, undated)

Doeksen, J. and van Wingerden, C. G. (1964). Notes on the activity of earthworms. 2. Observations on diapause in the earthworm *A. caliginosa. Jaarb. I.B.S.* 181–6.

Doerell, E. C. (1950). How do earthworms react to the application of minerals. *Deutsche Landwirtschaft. Presse,* **4,** 19.

Dotterweich, H. (1933). The function of storage of calcium by animals as a buffer reserve in the regulation of reaction. The calciferous glands of earthworms. *Pflügers Arch. ges. Physiol.* **232,** 263–86.

Dowdy, W. W. (1944). Influence of temperature on vertical migration. *Ecology of invertebrates inhabiting different soil types,* **25** (4) 449–60.

Dreidax, L. (1931). Investigations on the importance of earthworms for plant growth. *Arch. Pflanzenbau,* **7,** 413–67.

Drift, J. van der (1963). The influence of biocides on the soil fauna. *Neth. J. Pl. Path.* **69,** 188–99.

Dustman, E. H. and Stickel, L. F. (1966). Pesticide residues in the ecosystem. 'Pesticides and their effects on soils and water.' *Am. Soc. of Agronomy Spec. Publ.* **8,** 109–21.

Dutt, A. K. (1948). Earthworms and soil aggregation. *J. Am. Soc. Agron.* **40,** 407.

Duweini, A. K. and Ghabbour, S. I. (1965). Population density and biomass of earthworms in different types of Egyptian soils. *J. appl. Ecol.* **2,** 271–87.

Dzangaliev, A. D. and Belousova, N. K. (1969). Earthworm populations in irrigated orchards under various soil treatments. *Pedobiologia,* **9,** 103–5.

Eaton, T. H. Jr. (1942). Earthworms of the North-eastern United States. *J. Wash. Acad. Sci.* **32** (8) 242–9.

Eaton, T. H. and Chandler, R. F. (1942). The fauna of forest-humus layers in New York. *Mem. 247. Cornell Agr. Exp. Stn.* 26 pp.

Eberhardt, A. I. (1954). *Sarcophaga carnaria* als obligatorischer Regenwurm parasit. *Naturwissenschaften,* **41** (18) 436.

Edwards, C. A. (1965). Effects of pesticide residues on soil invertebrates and plants. *5th Symp. Brit. Ecol. Soc.* Blackwell, Oxford, pp. 239–61.

Edwards, C. A. (1970). Persistent pesticides in the environment. *Critical Reviews in Environmental Control.* Chem. Rubber Co., pp. 6–68.

Edwards, C. A. (1970). Effects of herbicides on the soil fauna. *Proc. 10th Weed Control Conf. 1970,* **3,** 1052–62.

Edwards, C. A. and Arnold, M. (1966). Effects of insecticides on soil fauna. *Rep. Rothamsted exp. Stn. for 1965,* pp. 195–6.

Edwards, C. A. and Dennis, E. B. (1960). Some effects of aldrin and DDT on the soil fauna of arable land. *Nature, Lond.* **188** (4572) 767.

Edwards, C. A., Dennis, E. B. and Empson, D. W. (1967). Pesticides and the soil fauna. 1. Effects of Aldrin and DDT in an arable field. *Ann. appl. Biol.* **59** (3) 11–22.

Edwards, C. A. and Heath, G. W. (1963). The role of soil animals in breakdown of leaf material. In *Soil Organisms,* J. Doeksen and van der Drift (eds.). North Holland Publishing Co., Amsterdam, pp. 76–80.

Edwards, C. A. and Lofty, J. R. (1969). Effects of cultivation on earthworm populations. *Rep. Rothamsted exp. Stn. for 1968,* 247–8.

Edwards, C. A. and Lofty, J. R. (1969). The influence of agricultural practice on soil micro-arthropod populations. In *The Soil Ecosystem.* Systematics Association publication No. 8. J. G. Sheals (ed.), pp. 237–47.

Edwards, C. A., Lofty, J. R. and Stafford, C. J. (1972). Insecticides and total soil fauna. *Rep. Rothamsted exp. Stn. for 1971* (in press).

Edwards, C. A. and Lofty, J. R. (1972). Effects of insecticides on soil invertebrates. *Rep. Rothamsted exp. Stn. for 1971* (in press).

Edwards, C. A., Reichle, D. E., Crossley, D. A. Jr. (1970). The Role of Soil Invertebrates in Turnover of Organic Matter and Nutrients. In *Ecological Studies, Analysis and Synthesis.* Springer-Verlag, Berlin, 147–172.

Edwards, C. A., Thompson, A. R. and Beynon, K. (1967). Some effects of chlorfenvinphos, an organophosphorus insecticide, on populations of soil animals. *Rev. Ecol. Biol. Sol.* **5** (2) 199–214.

Edwards, C. A., Whiting, A. E. and Heath, G. W. (1970). A mechanized washing method for separation of invertebrates from soil. *Pedobiologia,* **10** (5) 141–8.

El-Duweini, A. K. and Ghabour, S. I. (1965). Temperature relations of three Egyptian oligochaete species. *Oikos,* **16,** 9–15.

El-Duweini, A. K. and Ghabbour, S. I. (1965). Population density and biomass of earthworms in different types of Egyptian soils. *J. appl. Ecol.* **2,** 271–87.

Ellenby, C. (1945). Influence of earthworms on larval emergence in the potato root eelworm, *Heterodera rostochrensis* Wollenweber. *Ann. appl. Biol.* **31** (4) 332–9.

Escherich, K. (1911). *Termitenleben auf Ceylon Jena,* 263 pp.

Escritt, J. R. (1955). Calcium Arsenate for earthworm control. *J. Sports Turf Res. Inst.* **9** (31) 28–34.

Escritt, J. R. and Arthur, J. H. (1948). Earthworm control – a resumé of methods available. *J. Bd. Greenkeep. Res.* **7** (23) 49.

Evans, A. C. (1946). Distribution of numbers of segments in earthworms and its significance. *Nature, Lond.* **158,** 98.

Evans, A. C. (1947). Some earthworms from Iowa, including a description of a new species. *Ann. Mag. nat. Hist.* (11) **14,** 514.

Evans, A. C. (1947). Method of studying the burrowing activity of earthworms. *Ann. Mag. nat. Hist.* **11** (14) 643–50.

Evans, A. C. (1948). Some effects of earthworms on soil structure. *Ann. appl. Biol.* **35,** 1–13.

Evans, A. C. (1948). Relation of worms to soil fertility. *Discovery,* Norwich, **9** (3) 83–6.

Evans, A. C. (1948). Identity of earthworms stored by moles. *Proc. zool. Soc. Lond.* **118,** 1356–9.

Evans, A. C and Guild, W. J. Mc. L. (1947). Some notes on reproduction in British Earthworms. *Ann. Mag. nat. Hist.* 654.

Evans, A. C. and Guild, W. J. Mc. L. (1947). Cocoons of some British Lumbricidae. *Ann. Mag. nat. Hist.* 714–19.

Evans, A. C. and Guild, W. J. Mc. L. (1947). Studies on the relationships between earthworms and soil fertility. 1. Biological studies in the field. *Ann. appl. Biol.* **34** (3) 307–30.

Evans, A. C. and Guild, W. J. Mc. L. (1948). Studies on the relationships between earthworms and soil fertility. IV. On the life cycles of some British Lumbricidae. *Ann. appl. Biol.* **35** (4) 471–84.

Evans, A. C. and Guild, W. J. Mc. L. (1948). Studies on the relationships between earthworms and soil fertility. V. Field populations. *Ann. appl. Biol.* **35** (4) 485–93.

Feldkamp, J. (1924). Untersuchungen über die Geschlechtsmerkmale und die Begattung der Regenwürmer. *Zool. Jb.* (*Anat.*) **46,** 609–32.

Fenton, G. R. (1947). Ecological note on worms in forest soil. *J. Anim. Ecol.* **16,** 76–93.

Finck, A. (1952). Ökologische und Bodenkundliche Studien über die Leistungen der Regenwürmer für die Bodenfruchtbarkeit. *Z. PflErnähr. Düng.* **58,** 120–45.

Fleming, W. E. and Hadley, C. H. (1945). DDT ineffective for control of an exotic earthworm. *J. econ. Ent.* **38,** 411.

Fleming, W. E. and Hanley, I. M. (1950). A large scale test with DDT to control the Japanese beetle. *J. econ. Ent.* **43,** 586–90.

Ford, J. (1935). Soil communities in Central Europe. *J. Anim. Ecol.* **6,** 197–8.

Fox, C. J. S. (1964). The effects of five herbicides on the numbers of certain invertebrate animals in grassland soils. *Can. J. Pl. Sci.* **44,** 405–9.

Franz. H. and Leitenberger, L. (1948). Biological–chemical investigations into the formation of humus through soil animals. *Ost. zool. Z.* **1** (5) 498–518.

Gansen, P. S. van (1956). Les cellules chloragogenes des Lombriciens. *Bull. biol. Fr. Belg.* **90,** 335–56.

Gansen, P. S. van (1957). Histophysiologie du tube digestif d' *Eisenia foetida* (Sav.) region buccale, pharynx et glandes pharyngiennes. *Bull. biol. Fr. Belg.* **91,** 225–39.

Gansen, P. S. van (1958). Physiologie des cellules chloragogenes d'un lombricien. *Enzymologia,* **20,** 98–108.

Gansen, P. S. van (1962). Structures et functions du tube digestif du lombricien *Eisenia foetida* Savigny. Pub. Imp. Med. Sci. Bruxelles. 120 pp.

Garner, M. R. (1953). The preparation of latex casts of soil cavities for the study of the tunneling habits of animals. *Science,* **118,** 380–1.

Gast, J. (1937). Contrast between the soil profiles developed under pines and hardwood. *J. For.* **35,** 11–16.

Gates, G. E. (1929). The earthworm fauna of the United States. *Science,* **70,** 266–7.

Gates, G. E. (1949). Miscellanea megadrilogica. *Am. Nat.* **83,** 139–52.

Gates, G. E. (1954). On regenerative capacity of earthworms of the family Lumbricidae. *Am. Midl. Nat.* **50** (2) 414–19.

Gates, G. E. (1959). On a taxonomic puzzle and the classification of the earthworms. *Bull. Mus. comp. Zool. Harv.* **121,** 229–61.

Gates, G. E. (1961). Ecology of some earthworms with special reference to seasonal activity. *Am. Midl. Nat.* **66,** 61–86.

Gates, G. E. (1962). An exotic earthworm now domiciled in Louisiana. *Proc. Louisiana Acad. Sci.* **25,** 7–15.

Gates, G. E. (1963). Miscellanea Megadrilogica. VII. Greenhouse earthworms. *Proc. Biol. Soc. Wash.* **76,** 9–18.

Gates, G. E. (1966). Requiem for Megadrile Utopias. A contribution toward the understanding of the earthworm fauna of North America. *Proc. Biol. Soc. Wash.* **79,** 239–54.

Genov, T. (1963). Detection of the cysticercoid *Parieterotaenia paradoxa* (Rudolphi, 1802) (Dilepididae Fuhrmann, 1907) in *Allolobophora caliginosa* (Sav.) f. *trapezoides* (A. Dug.) (Lumbricidae). *Zool. Zh.* **42,** 1578–9.

Geoghegan, M. J. and Brain, R. C. (1948). Aggregate formation in soil. I. Influence of some bacterial polysaccharides on the binding of soil particles. *Biochem. J.* **43,** 5–13.

Gerard, B. M. (1960). The biology of certain British earthworms in relation to environmental conditions. Ph.D. thesis, London. 214 pp.

Gerard, B. M. (1964). *Synopses of the British Fauna.* (6) *Lumbricidae.* Linn. Soc. London. 58 pp.

Gerard, B. M. (1967). Factors affecting earthworms in pastures. *J. Anim. Ecol.* **36,** 235–52.

Gersch, M. (1954). Effect of carcinogenic hydrocarbons on the skin of earthworms. *Naturwissenschaften,* **41,** 337.

Ghabbour, S. I. (1966). Earthworms in agriculture: a modern evaluation. *Rev. Ecol. Biol. Soc.* **111** (2) 259–71.

Ghilarov, M. S. (1956). Significance of the soil fauna studies for the soil diagnostics. *6th Congr. Sci. Sol. Paris,* **3,** 130–44.

Ghilarov, M. S. (1956). Soil fauna investigation as a method in soil diagnostics. *Bull. Lab. Zool. 'Filipo Silvestri' Portici,* **33,** 574–85.

Ghilarov, M. S. (1963). On the interrelations between soil dwelling invertebrates and soil microorganisms. In *Soil Organisms,* J. Doeksen and J. van der Drift (eds). North Holland Publishing Co., Amsterdam. pp. 255–9.

Ghilarov, M. S. (1965). Zoological methods in soil diagnostics. In *'Nauka'*, Moscow, p. 278.

Ghilarov, M. S. and Byzova, J. B. (1961). Vlijanie Chimiceskich Obrabotok Lesa Na Pocuennuja Faunu. *Lesn. Ch-Vo.* **10,** 58–9.

Ghilarov, M. S. and Mamajev, B. M. (1963). Soil-inhabiting insects in irrigated regions of Uzbekistan. *Zashchita Rast of Vreidilelei Bodeznei,* **8,** 21–2.

Ghilarov, M. S. and Mamajev, B. M. (1966). Über die Ansiedlung von Regenwürmern in den artesisch bewässerten Oasen der Würste Kyst-Kum. *Pedobiologia,* **6,** 197–218.

Gish, C. D. (1970). Organochlorine insecticide residues in soils and soil invertebrates from agricultural land. *Pest. Mon. J.* **3** (4) 241–52.

Goffart, H. (1949). Die Wirkung neuer Insektiziden Mittel auf Regenwürmer. *Anz. f. Schädlingskunde,* **22,** 72–4.

Graff, O. (1953). Investigations in soil zoology with special reference to the terricole Oligochaeta. *Z. PflErnähr. Düng,* **61,** 72–7.

Graff, O. (1953). Die Regenwürmer Deutschlands. *Schrift. Forsch. Land. Braunschweig-Volk,* **7,** 81.

Graff, O. (1967). Translocation of nutrients into the subsoil through earthworm activity. *Landw. Forsch.* **20,** 117–27.

Graff, O. (1969). Regenwurmtätigkeit in Ackerböden unter verschiedenem Bedeckungsmaterial, gemessen an der Lösungsablage. *Pedobiologia,* **9** (1–2), 120–8.

Graff, O. (1972). Stickstoff, Phosphor und Kalium in der Regenwurmlosung auf der Wiesenversuchs-fläche des Sollingprojektes. *Proc. IVth. Symp. Soil Zool.* (in press).

Graff, O. and Satchell, J. E. (1967). *Progress in Soil Biology Pub.* North-Holland Publishing Company, Amsterdam. 656 pp.

Grant, W. C. (1955). Studies on moisture relationships in earthworms. *Ecology,* **36** (3) 400–7.

Grant, W. C. (1955). Temperature relationships in the megascolecid earthworm, *Pheretima hupeiensis. Ecology,* **36** (3) 412–17.

Grant, W. C. (1956). An ecological study of the peregrine earthworm, *Pheretima hupeiensis* in the Eastern United States. *Ecology,* **37,** (4) 648–58.

Grassi, B. and Rovelli, G. (1892). Recherche embriologiche sui Cestodi. *Att. Asc. Catania,* **4,** 15–108.

Gray, J. and Lissmann, H. W. (1938). Studies on Animal Locomotion. VII. Locomotory reflexes in the earthworm. *J. exp. Biol.* **15** (4) 506–17.

Greenwood, D. E. (1945). Wireworm investigations. *Conn. agric. Exp. Stn. Bull.* **488,** 344–7.

Griffiths, D. C., Raw, F. and Lofty, J. R. (1967). The effects on soil fauna of insecticides tested against wireworms (*Agriotes* spp.) in wheat. *Ann. appl. Biol.* **60,** 479–90.

Grigor'eva, T. G. (1952). The action of BHC introduced into the soil on the soil fauna. *Dokl. vsesoyuz. Akad. selkhoz Nauk. Lenina.* **17,** 16–20; summary *Rev. appl. Ent.* (A) **41,** 336.

Grove, A. J. and Newell, G. E. (1962). *Animal Biology.* Univ. Tutorial Press, London. 820 pp.

Guild, W. F. Mc. L. (1948). Effect of soil type on populations. *Ann. appl. Biol.* **35** (2) 181–92.

Guild, W. J. Mc. L. (1951). Earthworms in Agriculture. *Scot. agric.* **30** (4) 220–3.

Guild, W. J. Mc. L. (1951). The distribution and population density of earthworms (Lumbricidae) in Scottish pasture fields. *J. Anim. Ecol.* **20** (1) 88–97.

Guild, W. J. Mc. L. (1952). Variation in earthworm numbers within field populations. *J. Anim. Ecol.* **21** (2) 169.

Guild, W. J. Mc. L. (1952a). The Lumbricidae in upland areas. 11. Population variation on hill pasture. *Ann. Mag. nat. Hist.* **12** (5) 286–92.

Guild, W. J. Mc. L. (1955). Earthworms and soil structure. *In. Soil Zoology,* D. K. Mc. E. Kevan (ed.). Butterworths, London. pp. 83–98.

Gunthart, E. (1947). Die Bekämpfung der Engerlinge mit Hexachlorocyclohexan – Präparation. *Mitt. Schweiz Ent. Ges.* **20,** 409–50.

Gurianova, O. Z. (1940). Effect of earthworms and of organic fertilisers on structure formation in chernozem soils. *Pedology,* **4,** 99–108.

Hamblyn, C. J. and Dingwall, A. R. (1945). Earthworms. *N.Z. Jl. Agric.* **71,** 55–8.

Hanel, E. (1904). Ein Beitrag zur 'Psychologie' der Regenwürmer. *Z. allg. Physiol.* **4,** 244–58.

Harmsen, G. and van Schreven, D. (1955). Mineralisation or organic nitrogen in soil. *Adv. Agron.* **7,** 299–398.

Hasenbein, G. (1951). A pregnancy test on earthworms. *Arch. Gynakol.* **181,** 5–28.

Haswell, W. A. and Hill, J. P. (1894). A proliferating cystic parasite of the earthworms. *Proc. Linn. Soc. N.S. Wales,* **8** (2) 365–76.

Heath, G. W. (1962). The influence of ley management on earthworm populations. *J. Br. Grassld Soc.* **17** (4) 237–44.

Heath, G. W. (1965). The part played by animals in soil formation. In *Experimental Pedology,* E. G. Hallsworth and D. V. Crawford (eds). Butterworths, London. pp. 236–43.

Heath, G. W., Arnold, M. K. and Edwards, C. A. (1966). Studies in leaf litter breakdown. 1. Breakdown rates among leaves of different species. *Pedobiologia*, **6**, 1–12.

Heath, G. W. and King, H. G. C. (1964). The palatability of litter to soil fauna. *Proc. VIII Int. Congr. Soil Sci. Bucharest*, pp. 979–86.

Heck, L. von. (1920). Über die Bildung einer Assoziation beim Regenwurm auf Grund von Dressurversuchen. *Lotos Naturwiss. Z.*, **68**, 168–89.

Heimburger, H. V. (1924). Reactions of earthworms to temperature and atmospheric humidity. *Ecology*, **5**, 276–83.

Hensen, V. (1877). Die Tätigkeit des Regenwurms (*L. terrestris*) für die Fruchtbarkeit des Erdbodens. *Z. wiss. Zool.* **28**, 354–64.

Herlant-Meewis, H. (1956). Croissance et reproduction du Lombricien, *Eisenia foetida* (Sav.). *Ann. Sci. nat. Zool. Biol. Anim.*, **18**, 185–98.

Hess, W. N. (1924). Reactions to light in the earthworm, *Lumbricus terrestris*. *J. Morph.* **39**, 515–42.

Hess, W. N. (1925). Nervous system of the earthworm, *Lumbricus terrestris* L. *J. Morph.* **40**, 235–60.

Hess, W. N. (1925a). Photoreceptors of *Lumbricus terrestris*, with special reference to their distribution. *J. Morph.* **41**, 235–60.

Heungens, A. (1966). Bestrijding van Regenwormen in Sparregrond en in vitro. *Med. Rijksfak. Landb. w. Sch. Gent*, **31**, 329–42.

Heungens, A. (1969). L'influence de la fumure et des pesticides aldrine, carbaryl et DBCP sur la faune du sol dans la culture des azalées. *Rev. Ecol. Biol. Soc.* **6** (2) 131–45.

Heungens, A. (1969). The physical decomposition of pine litter by earthworms. *Pl. Soil*, **31** (1) 22–30.

Hirst, J. M., Storey, Ward, W. C. and Wilcox, H. G. (1955). The origin of apple scab epidemics in the Wisbech area in 1953 and 1954. *Pl. Path.* **4**, 91.

Hobmaier, A. and Hobmaier, M. (1929). Die Entwicklung der Larve des Lungenwurmes *Metastrongylus elongatus* (*Strongylus paradoscus*) des Schweines und ihr Invasionsweg. *Munch. Tierärzt. Wschr.* **80**, 365–9.

Hoeksema, K. J., Jongerious, A. and K. van der Meer (1956). On the influence of earthworms on the soil structure in mulched orchards. *Boor en Spade*, **8**, 183–201.

Hoffman, J. A. and Purdy, L. H. (1964). Germination of dwarf bunt (*Tilletia controversa*) teliospores after ingestion by earthworms. *Phytopathology*, **54**, 878–9.

Hogben, L. and Kirk, R. L. (1944). Body temperature of worms in moist and dry air. *Proc. Roy. Soc. Lond.* **132**b, (868) 239–52.

Hopkins, A. R. and Kirk, V. M. (1957). Effects of several insecticides on the English red worm. *J. econ. Ent.* **50** (5) 699–700.

Hopp, H. (1946). Earthworms fight erosion too. *Soil Conserv.* **11,** 252–4.

Hopp, H. (1947). The ecology of earthworms in cropland. *Soil. Sci. Soc. Amer. Proc.* **12,** 503–7.

Hopp, H. and Hopkins, H. T. (1946). Earthworms as a factor in the formation of water-stable aggregates. *J. Soil Water Conserv.* **1,** 11–13.

Hopp, H. and Hopkins, H. T. (1946b). The effect of cropping systems on the winter populations of earthworms. *J. Soil Water Conserv.* **1** (1) 85–8, 98.

Hopp, H. and Slater, C. S. (1948). Influence of earthworms on soil productivity. *Soil Sci.* **66,** 421–8.

Hopp, H. and Slater, C. S. (1949). The effect of earthworms on the productivity of agricultural soil. *J. agric. Res.* **78,** 325–39.

Howell, C. D. (1939). Nervous responses of *Pheretima* to light. *J. exp. Zool.* **81,** 231–59.

Hoy, H. M. (1955). Toxicity of some hydrocarbon insecticides to earthworms. *N.Z. Jl Sci. Technol.* (A) **37** (4) 367–72.

Hubl, H. (1953). Die inkretorischen Zellelemente im Gehirn der Lumbriciden. *Arch. EntwMech. Org.* **146,** 421–32.

Hubl, H. (1956). Uber die Beziehungen der Neurosekretion zum Regenerations Geschehen bei Lumbriciden nebst Beschreibung eines neuartigen neurosekretorischen Zelltyps in Unterschlundganglion. *Arch. EntwMech. Org.* **149,** 73–87.

Hunt, L. B. (1965). Kinetics of pesticide poisoning in Dutch Elm Disease control. *U.S. Fish Wildl. Serv. Circ.* **226,** 12–13.

Hunt, L. B. and Sacho, R. J. (1969). Response of robins to DDT and methoxychlor. *J. Wildlife Manage,* **33,** 267–72.

Hutchinson, S. A. and Kamel, M. (1956). The effect of earthworms on the dispersal of soil fungi. *J. Soil Sci.* **7** (2) 213–18.

Hyche, L. L. (1956). Control of mites infesting earthworm beds. *J. econ. Ent.* **49,** 409–10.

Hyman, L. H. (1940). Aspects of regeneration in Annelids. *Am. Nat.* **74,** 513–27.

Inoue, T. and Kondo, K. (1962). Susceptibility of *Branchiura sowerbyi, Limrodrilus socialis* and *L. willeyi* for several agricultural chemicals. *Botyu-bagaku (Japan),* **27,** 97–9.

Jacks, G. V. (1963). The biological nature of soil productivity. *Soils & Fert.* **26** (3) 147–50.

Jacob, A. and Wiegland, K. (1952). Transformations of the mineral nitrogen of fertilisers in the soil. *Z. PflErnähr. Düng.* **59,** 48–60.

Jeanson-Luusinang, C. (1961). Sur une methode d'étude du comportement de la fauna du sol et de sa contribution to pédogenese. *C.R. Acad. Sci.* **253,** 2571–3.

Jeanson-Luusinang, C. (1963). Action des Lombricides sur la microflore totale. In *Soil Organisms,* J. Doeksen and J. van der Drift (eds). North-Holland Pub. Co., Amsterdam. pp. 260–5.

Jefferson, P. (1955). Studies on the earthworms of turf. A. The earthworms of experimental turf plots. *J. Sports Turf Res. Inst.* **9** (31) 6–27.

Jefferson, P. (1956). Studies on the earthworms of turf. B. Earthworms and soil. *J. Sports Turf Res. Inst.* **9** (31) 6–27.

Joachim, A. W. R. and Panditesekera, D. G. (1948). Soil fertility studies. IV. Investigations on crumb structure on stability of Local soils. *Trop. Agric.* **104,** 119–39.

Johnson, M. L. (1942). The respiratory function of the haemoglobin of the earthworm. *J. exp. Biol.* **18** (3) 266–77.

Johnstone-Wallace, D. B. (1937). The influence of wild white clover on the seasonal production and chemical composition of pasture herbage and upon soil temperatures. Soil moistures and erosion control. *4th Int. Grassl. Congr. Rep.* 188–96.

Joshi, N. V. and Kelkar, B. V. (1952). The role of earthworms in soil fertility. *Indian J. agric. Sci.* **22,** 189–96.

Kahsnitz, H. G. (1922). Investigations on the influence of earthworms on soil and plant. *Bot. Arch.* **1,** 315–51.

Kalmus, H. (1955). On the colour forms of *Allolobophora chlorotica* Sav. *Ann. Mag. nat. Hist.* **12** (8) 795.

Karmanova, E. M. (1959). Biology of the nematode *Hystrichis tricolor* Dujardin, 1845, and some data on epizootics in ducks: A histochemical study. *J. Univ. Bombay,* **30B,** 113–25.

Karmanova, E. M. (1963). Interpretation of the developmental cycle in *Dioctophyme renale. Med. Parazitol. Paraziter, Boleznii* **32,** 331–4.

Keilin, D. (1915). Recherches sur les larves de Diptères cyclorrhaphes. *Bull. scient. Fr. Belg.* **47,** 15–198.

Keilin, D. (1925). Parasitic autotomy of the host as a mode of liberation of coelomic parasites from the body of the earthworm. *Parasitology,* **17,** 70–2.

Kelsey, J. M. and Arlidge, G. Z. (1968). Effects of Isobenzan on soil fauna and soil structure. *N.Z. Jl. agric. Res.* **11,** 245–60.

Kevan, D. K. Mc. E. (1955). Soil Zoology, (ed.). Butterworths, London, pp. 23–8, 452–88.

Khambata, S. R. and Bhatt, J. V. (1957). A contribution to the study of the intestinal microflora of Indian Earthworms. *Arch. Mikrobiol.* **28,** 69–80.

King, H. G. C. and Heath, G. W. (1967). The chemical analysis of small samples of leaf material and the relationship between the disappearance and composition of leaves. *Pedobiologia*, **7**, 192–7.

Kirberger, C. (1953). Untersuchungen über die Temperaturabhängigkeit von Lebensprozessen bei verschiedenen Wirbellosen. *Z. vergl. Physiol.* **35**, 175–98.

Knop, J. (1926). Bakterien und Bacteroiden bei oligochäten. *Z. Morph. Ökol. Tiere*, **6**.

Kobatake, M. (1954). The antibacterial substance extracted from lower animals. 1. The earthworm. *Kekkabu* (*Tuberculosis*), **29**, 60–3.

Kollmannsperger, G. (1934). The Oligochaeta of the Bellinchen Region. Inaugural dissertation. Dillingen (Saargebiet).

Kollmannsperger, F. (1955). Über Rhythmen bei Lumbriciden. *Decheniana*, **180**, 81–92.

Kollmannsperger, F. (1956). Lumbricidae of humid and arid regions and their effect on soil fertility. *VI Congr. Int. Sci. Sol. Rapp.* C., 293–7.

Korschelt, E. (1914). Über Transplantationsversuche, Ruhezustände und Lebensdauer der Lumbriciden. *Zool. Anz.* **43**, 537–55.

Kozlovskaya, L. S. and Zaguralskaya, L. M. (1966). Relationships between earthworms and microbes in W. Siberia. *Pedobiologia*, **6**, 244–57.

Kozlovskaya, L. S. and Zhdannikova, E. N. (1961). Joint action of earthworms and microflora in forest soils. *Dokl. Akad. Nauk. SSSR* **139**, 470–73.

Kring, J. B. (1969). Mortality of the earthworm *Lumbricus terrestris* L. following soil applications of insecticides to a tobacco field. *J. econ. Ent.* **62** (4) 963.

Krivanek, J. O. (1956). Habit formation in the earthworm, *Lumbricus terrestris*. *Physiol. Zool.*, **29**, 241–50.

Krüger, F. (1952). Über die Beziehung des Sauerstoffverbauchs zum Gewicht bei *Eisenia foetida* Sav. *Z. vergl. Physiol.* **34**, 1–5.

Kubiena, W. L. (1953). *Bestimmungsbuch und Systematik der Böden Europas*. Stuttgart, p. 392.

Kubiena, W. L. (1955). Animal activity in soils as a decisive factor in establishment of humus forms, in *Soil Zoology*, Kevan (ed.). Butterworths, London, pp. 73–82.

Kuhnelt, W. (1961). *Soil Biology*. Faber and Faber, London. pp. 397.

Kurcheva, G. F. (1960). The role of invertebrates in the decomposition of the oak leaf litter. *Pocvovedenie* (4) 16–23.

Ladell, W. R. S. (1936). A new apparatus for separating insects and other arthropods from the soil. *Ann. appl. Biol.* **23**, 862–79.

Lakhani, K. H. and Satchell, J. E. Production by *Lumbricus terrestris* (L.). *J. Anim. Ecol.* **39,** 473–492.

Lan an der, H. and Aspöck, H. (1962). Zur Wirkung von Sevin auf Regenwürmer. *Anz. Schädlingsk.* **35,** 180–2.

Lauer, A. R. (1929). Orientation in the earthworm. *Ohio. J. Sci.,* **29,** 179.

Laverack, M. S. (1960). Tactile and chemical perception in earthworms. 1. Responses to touch, sodium chloride, quinine and sugars. *Comp. Biochem. Physiol.* **1,** 155–63.

Laverack, M. S. (1960). The identity of the porphyrin pigments of the integument of earthworms. *Comp. Biochem. Physiol.* **1** (4) 259–66.

Laverack, M. S. (1961). Tactile and chemical perception in earthworms. 11. Responses to acid pH solutions. *Comp. Biochem. Physiol.* **2** (1) 22–34.

Laverack, M. S. (1961). The effect of temperature changes on the spontaneous nervous activity of the isolated nerve cord of *Lumbricus terrestris. Comp. Biochem. Physiol.* **3** (2) 136–40.

Laverack, M. S. (1963). *The Physiology of Earthworms.* Pergamon Press, London. 206 pp.

Lawrence, R. D. and Millar, H. R. (1945). Protein content of earthworms. *Nature, Lond.* **155** (3939) 517.

Lee, K. E. (1951). Role of earthworms in New Zealand soil. *Tuatara,* **4** (1) 22–7.

Lee, K. E. (1959). A key for the identification of New Zealand earthworms. *Tuatara,* **8** (1) 13–60.

Legg, D. C. (1968). Comparison of various worm-killing chemicals. *J. Sports Turf Res. Inst.* **44,** 47–8.

Lesser, E. J. (1910). Chemische Prozesse bei Regenwürmern. 3. Über anoxybiotische Zersetzung des Glykogens. *Z. Biol.* **50.**

Lidgate, H. J. (1966). Earthworm control with chlordane. *J. Sports Turf Res. Inst.* **42,** 5–8.

Lindquist, B. (1941). Investigations on the significance of some Scandinavian earthworms in decomposition of leaf litter and the structure of mull soil. *Svensk Skogs v Fören Tidskr.* **39** (3) 179–242.

Lipa, J. J. (1958). Effect on earthworm and Diptera populations of BHC dust applied to soil. *Nature, Lond.* **181,** 863.

Ljungström, P. O. (1964). Ekologin hos daggmaskar i Stockholmstrakten. *Fält. Biologen.* **2.**

Ljungström, P. O. and Reinecke, A. J. (1969). Ecology and natural history of the microchaelid earthworms of South Africa. 4. Studies on influence of earthworms upon the soil and the parasitological question. *Pedobiologia,* **9** (1–2) 152–7.

Lofty, J. R. (1972). The effects of gamma radiation on earthworms. *Pedobiologia* (in press).

Long, W. H., Anderson, H. L. and Isa, A. L. (1967). Sugarcane growth responses to chlordane and microarthropods, and effects of chlordane on soil fauna. *J. econ. Ent.* **60,** 623–9.

Low, A. J. (1955). Improvements in the structural state of soils under leys. *J. Soil Sci.* **6,** 179–99.

Luckman, W. H. and Decker, G. C. (1960). A 5-year report of observations in the Japanese beetle control area at Sheldon, Illinois. *J. econ. Ent.* **53,** 821–7.

Lukose, J. (1960). A note on an association between two adult earthworms. *Curr. Sci.* **29,** 106–7.

Lund, E. E., Wehr, E. E. and Ellis, D. J. (1963). Role of earthworms in transmission of *Heterakis* and *Histomonas* to turkeys and chickens. *J. Parasit.* **49** (5) 50.

Lunt, H. A. and Jacobson, G. M. (1944). The chemical composition of earthworm casts. *Soil Sci.* **58,** 367.

McLeod, J. H. (1954). Note on a staphylinid (Coleoptera) predator of earthworms. *Canad. Ent.* **86,** 236.

Madge, D. S. (1966). How leaf litter disappears. *New Scientist,* **32,** 113–15.

Madge, D. S. (1969). Field and laboratory studies on the activities of two species of tropical earthworms. *Pedobiologia,* **9,** 188–214.

Magalhaes, P. S. (1892). Notes d'helminthologie brésilienne. *Bull. Soc. Zool. France,* **17,** 145–6.

Maldague, M. and Couture, G. (1972). Utilisation de litières radioactives par *Lumbricus terrestris. Proc. 4th Int. Congr. Soil Zool.* (in press).

Mamytov, A. (1953). The effect of earthworms on the water stability of mountain-valley serozem soils. *Pochvovederie,* **8,** 58–60.

Mangold, O. (1951). Experiments in analysis of the chemical senses of earthworms. 1. Methods and procedure for leaves of plants. *Zool. Jb.* (*Physiol.*) **62,** 441–512.

Mangold, O. (1953). Experimente zur Analyse des chemischen Sinns des Regenwurms. 2. Versuche mit Chinin, Säuren und Süsstoffen. *Zool. Jb. Abt. Allgem. Zool. Physiol. Tiere.* **63,** 501–57.

Marapao, B. P. (1959). The effect of nervous tissue extracts on neurosecretion in the earthworm *Lumbricus terrestris. Catholic U. Amer. Biol. Stud.* **55,** 1–34.

Marshall, V. G. (1972). Effects of soil arthropods and earthworms on the growth of Black Spruce. *Proc. 4th Int. Congr. Soil Zool.* (in press).

Martin, A. W. (1957). Recent advances in knowledge of invertebrate renal function. In *Invertebrate Physiology*, B. T. Scheer (ed.) University of Oregon Pub., pp. 247–76.

Meggitt, F. J. (1914). On the anatomy of a fowl tapeworm, *Amoebotaenia sphenoides v. Linstow. Parasitology*, **7**, 262–77.

Mellanby, K. (1961). Earthworms and the soil. *Countryside*, **14** (4) 1.

Merker, E. and Braunig, G. (1927). Die Empfindlichkeit feuchthäutiger Tiere im Lichte. 3. Die Atemnot feuchthäutiger Tiere in Licht der Quarzquecksiblerlampe. *Zool. Jb. Abt. Allgem. Zool. Physiol. Tiere*, **43,** 275–338.

Meyer, L. (1943). Experimental study of macrobiological effects on humus and soil formation. *Bodenk. u. PflErnähr.* **29** (74) 119–40.

Michaelsen, W. (1910). Die Oligochätenfauna der vorderindischceylonischen Region. *Abh. Naturw. Hamburg.* **19.**

Michaelsen, W. (1921). Zur Stammesgeschichte und Systematik der Oligochäten, insbesondere der Lumbriculiden. *Arch. Naturgesch.* **86.**

Michaelsen, W. (1922). Die Verbreitung der Oligochäten in Lichte der Wegener'schen Theorie der Kontinentverschiebung. *Verh. Ver. naturs. Unterh. Hamburg*, **3,** 29.

Michaelsen, W. (1926). *Pelodrilus bureschi*, ein Süsswasser-Höhlenoligochät aus Bulgarien. *Arb. Bulgar. Naturf. Ges.* **12,** 57–66.

Michaelsen, W. (1903). *Die geographische Verbreitung der Oligochaeten.* Berlin, 183 pp.

Michon, J. (1949). Influence of desiccation on diapause in Lumbricids. *C.r. hebd. Séanc. Acad. Sc., Paris*, **228** (18) 1455–6.

Michon, J. (1951). Supernumerary regeneration in *A. terrestris* f. *typica. C.r. hebd. Séanc. Acad. Sci., Paris*, **232,** 1449–51.

Michon, J. (1954). Influence de l'isolement à partir de la maturité sexuelle sur la biologie des *Lumbricidae. C.r. hebd. Séanc. Acad. Sci., Paris*, **238,** 2457–8.

Michon, J. (1957). Contribution experimentale á étude de la biologie des Lumbricidae. *Année Biol.* **33** (7–8) 367–76.

Miles, H. B. (1963). Soil protozoa and earthworm nutrition. *Soil Sci.* **95,** 407–9.

Miles, H. B. (1963). Heat-death temperature in *Allolobophora terrestris* f. *longa* and *Eisenia foetida. Nature, Lond.* **199,** 826.

Millott, N. (1944). The visceral nerves of the earthworm. 3. Nerves controlling secretion of protease in the anterior intestine. *Proc. R. Soc.* **132,** 200–12.

Moment, G. B. (1953). The relation of body level, temperature and nutrition to regenerative growth. *Physiol. Zool.* **26,** 108–17.

Moment, G. B. (1953). A theory of growth limitation. *Am. Nat.* **88** (834) 139–53.

Mönnig, H. O. (1927). The anatomy and life history of the fowl tapeworm *Amoebotaenia spheroides. Report of the Director of Veterinary Education and Research,* **11–12,** 199–206.

Moore, B. (1922). Earthworms and soil reaction. *Ecology,* **3,** 347–8.

Moore, A. R. (1923). Muscle tension and reflexes in the earthworm. *J. gen. Physiol.* **5,** 327–33.

Morris, H. M. (1922). Insect and other invertebrate fauna of arable land at Rothamsted. *Ann. appl. Biol.* **9** (3–4) 282–305.

Morrison, F. O. (1950). The toxicity of BHC to certain microorganisms. Earthworms and arthropods. *Ontario Ent. Soc. Ann. Rep.* **80,** 50–7.

Mozgovoy, A. A. (1952). The biology of *Porrocaecum crassum,* a nematode of aquatic birds. *Trudy gelmint. labor.* **6,** 114–25.

Muldal, S. (1949). Cytotaxonomy of British earthworms. *Proc. Linn. Soc. Lond.* **161,** 116–18.

Müller, G. (1965). *Bodenbiologie.* Verlag VEB Gustav Fischer, Jena. 889 pp.

Müller, P. E. (1878). Nogle Undersøgelser af Skovjord. *Tidsskr. Landøko,* **4,** 259–83.

Müller, P. E. (1884). Studier over Skovjord. 11. Om Muld og Mor i Egeskove og paa Heder. *Tidsskr. Skovbrug,* **7,** 1–232.

Müller, P. E. (1950). Forest-soil studies, a contribution to silvicultural theory. III. On compacted ground deficient in mull, especially in beach forests. *Damsk Skovfören. Tidsskr.* **1,** 10–61.

Murchie, W. R. (1955). A contribution on the natural history of *Allolobophora minima,* Muldal. *Ohio J. Sci.* **55** (4) 241–4.

Murchie, W. R. (1956). Survey of the Michigan earthworm fauna. *Mich. Acad. Sci. Arts. Let.* **151,** 53–72.

Murchie, W. R. (1958). Biology of the oligochaete *Eisenia rosea* (Savigny) in an upland forest soil of Southern Michigan. *Am. Midl. Nat.* **66** (1) 113–31.

Murchie, W. R. (1958). A new megascolecid Earthworm from Michigan with notes on its Biology. *Ohio J. Sci.* **58** (5) 270–2.

Murchie, W. R. (1959). Redescription of *Allolobophora muldali* Omodeo. *Ohio J. Sci.* **59** (6) 229–32.

Murchie, W. R. (1960). Biology of the oligochaete *Bimastos zeteki* Smith and Gittins (Lumbricidae) in Northern Michigan. *Am. Midl. Nat.* **64** (1) 194–215.

Murchie, W. R. (1961). A new diplocardian earthworm from Illinois. *Ohio J. Sci.* **61** (6) 367–71.

Murchie, W. R. (1961). A new species of Diplocardia from Florida. *Ohio J. Sci.* **61** (3) 175–7.

Murchie, W. R. (1963). Description of a new diplocardian earthworm, *Diplocardia longiseta*. *Ohio J. Sci.* **63** (1) 15–18.

Murchie, W. R. (1965). *Diplocardia gatesi*, a new earthworm from North Carolina. *Ohio J. Sci.* **65** (4) 208–11.

Nakamura, Y. (1968). Population density and biomass of the terrestrial earthworm in the grasslands of three different soil types near Sapporo. *Jap. J. appl. Ent. Zool.* **11,** 164–8.

Nakamura, Y. (1968). Studies on the ecology of terrestrial Oligochaetae. 1. Seasonal variation in the population density of earthworms in alluvial soil grassland in Sapporo, Hokkaido. *Jap. J. appl. Ent. Zool.* **3** (2) 89–95.

Needham, A. E. (1957). Components of nitrogenous excreta in the earthworms *L. terrestris* and *E. foetida*. *J. exp. Biol.* **34** (4) 425–46.

Needham, A. E. (1962). Distribution of arginase activity along the body of earthworms. *Comp. Biochem. Physiol.* **5,** 69–82.

Nelson, J. M. and Satchell, J. E. (1962). The extraction of Lumbricidae from soil with special reference to the hand-sorting method. *Progress in Soil Zoology*. P. Murphy (ed.). Butterworths, London, pp. 294–9.

Newell, G. E. (1950). The role of the coelomic fluid in the movements of earthworms. *J. exp. Biol.* **21** (1) 110–21.

Nielson, R. L. (1951). Effect of soil minerals on earthworms. *N.Z. Jl. Agric.* **83,** 433–5.

Nielson, R. L. (1952). Earthworms and soil fertility. *N.Z. Grassl. Assoc. Proc.* 158–67.

Nielsen, C. O. (1953). Studies on Enchytraeidae. 1. A technique for extracting Enchytraeidae from soil samples. *Oikos*, **4** (2) 187–96.

Nielson, R. L. (1953). Recent research work. Earthworms. *N.Z. Jl. Agric.* **86,** 374.

Nijhawan, S. D. and Kanwar, J. S. (1952). Physiochemical properties of earthworm castings and their effect on the productivity of soil. *Indian J. agric. Sci.* **22,** 357–73.

Nye, P. H. (1955). Some soil-forming processes in the humid tropics. IV. The action of soil fauna. *J. Soil Sci.* **6,** 78.

Ogg, W. G. and Nicol, H. (1945). Balanced manuring. *Scot. J. agric.* **25** (2) 76–83.

Oldham, C. (1915). *Testacella scutulum* in Hertfordshire. *Trans. Herts nat. Hist. Soc.* **15,** 193–4.

Olson, H. W. (1928). The earthworms of Ohio. *Ohio biol. Surv. Bull.* **17,** 47–90.

Omodeo, P. (1952). Cariologia Dei Lumbricadae. *Inst. Biol. Zool. Gen. Univ. Siena*, **4,** 173–275.

Omodeo, P. (1958). La réserve naturelle integralè du Mont Nimba. I. Oligochètes. *Mem. Inst. fr. Afr. noire.* **53,** 9–10.

Otanes, F. G. and Sison, P. L. (1947). Pests of Rice. *Philip. J. agric.* **13,** 36–88.

Parker, G. H. and Parshley, H. M. (1911). The reactions of earthworms to dry and moist surfaces. *J. exp. Zool.* **11,** 361–3.

Parle, J. N. (1959). Activities of micro-organisms in soil and influence of these on soil fauna. *Ph.D. Thesis, Lond.*

Parle, J. N. (1963). Micro-organisms in the intestines of earthworms. *J. gen. Microbiol.* **31,** 1–13.

Parle, J. N. (1963). A microbiological study of earthworm casts. *J. gen. Microbiol.* **13,** 13–23.

Patel, H. K. (1960). Earthworms in tobacco nurseries and their control. *Indian Tobacco,* **10** (1) 56.

Patel, H. K. and Patel, R. M. (1959). Preliminary observations on the control of earthworms by soapdust (*Sapindus laurifolius* Vahl) extract. *Indian J. Ent.* **21,** 251–5.

Peachey, J. E. (1963). Studies on the Enchytraeidae (Oligochaeta) of moorland soil. *Pedobiologia,* **2,** 81–95.

Peredel'sky, A. A. (1960a). Effect of earthworms and wireworms on absorption by plants of the radioactive isotopes Ca^{45} and Sr^{90} from soil. *Dokl. Akad. Nauk.* **134,** 1450–2.

Peredel'sky, A. A. (1960b). Dispersion of radioactive isotopes in the soil by earthworms. *Dokl. Akad. Nauk.* **135,** 185–8.

Peredel'sky, A. A., Poryadkova, N. A. and Rodionova, L. Z. (1957). The role of earthworms in purification of soil contaminated with radioactive isotopes. *Dokl. Akad. Nauk.* **115** (4) 809–12.

Parel, T. S. and Sokolov, D. F. (1964). Quantitative evaluation of the participation of the earthworm *Lumbricus terrestris* Linné (Lumbricidae – Oligochaeta) in the transformation of forest litter. *Zool. Zh.* **53,** 1618–25.

Perel, T. S., Karpachevskii, L. O. and Yegorova, S. V. (1966). Experiments for studying the effect of earthworms on the litter horizon of forest soils. *Pedobiologia,* **6,** 269–76.

Petrov, B. C. (1946). The active reaction of soil (pH) as a factor in the distribution of earthworms. *Zool. Jour.* **25** (1), 107–10.

Phillips, E. F. (1923). Earthworms, plants and soil reactions. *Ecology,* **4,** 89.

Polivka, J. B. (1951). Effect of insecticides upon earthworm populations. *Ohio J. Sci.* **51,** 195–6.

Polivka, J. B. (1953). More about the effect of insecticides on earthworm populations. *Unpublished mimeo. Ohio Acad. Sci.* 10 pp.

Pomerat, G. M. and Zarrow, M. T. (1936). The effect of temperature on the respiration of the earthworm. *Proc. natn. Acad. Sci. U.S.A.* **22,** 270–2.

Ponomareva, S. I. (1950). The role of earthworms in the creation of a stable structure in ley rotations. *Pochvovedenie*, 476–86.

Ponomareva, S. I. (1952). The importance of biological factors in increasing the fertility of sod-podzolic soils. *Z. PflErnähr. Düng.* **97,** 205–15.

Ponomareva, S. I. (1953). The influence of the activity of earthworms on the creation of a stable structure in a sod-podolized soil. *Trudy pochv. Inst. Dokuchaeva*, **41,** 304–78.

Ponomareva, S. I. (1962). Soil macro and micro-organisms and their role in increasing fertility. *Vtoraya Zoologischeskaya Konferenciya Litovskoi SSR*, 97–99.

Powers, W. L. and Bollen, W. B. (1935). The chemical and biological nature of certain forest soils. *Soil Sci.* **40,** 321–9.

Prabhoo, N. R. (1960). Studies on Indian Enchytraeidae (Oligochaeta: Annelida). Description of three new species. *J. Zool. Soc. India.* **12** (2) 125–32.

Prosser, C. L. (1935). Impulses in the segmental nerves of the earthworm. *J. exp. Biol.* **12,** 95–104.

Puh. P. C. (1941). Beneficial influence of earthworms on some chemical properties of the soil. *Contr. biol. Lab. sci. Soc. China*, **15,** 147–55.

Puttarudriah, M. and Sastry, K. S. S. (1961). A preliminary study of earthworm damage to crop growth. *Mysore Agric. J.* **36,** 2–11.

Raffy, A. (1930). La respiration des vers de terre dans l'eau. Action de la teneur en oxygêne et de la lumiéu sur l'intensité de la respiration pendant l'immersion. *C.r. hebd. Séanc. Acad. Sci. Paris*, **105,** 862–4.

Ragg, J. M. and Ball, D. F. (1964). Soils of the ultra-basic rocks of the Island of Rhum. *J. Soil Sci.* **15** (1) 124–34.

Ralph, C. L. (1957). Persistent rhythms of activity and O_2 consumption in the earthworm. *Physiol. Zoöl.* **30,** 41–55.

Ramsay, J. A. (1949). Osmotic relations of worms. *J. exp. Biol.* **26** (1) 65–75.

Raw, F. (1959). Estimating earthworm populations by using formalin. *Nature, Lond.* **184,** 1661.

Raw, F. (1960). Observations on the effect of hexoestrol on earthworms and other soil invertebrates. *J. agric. Sci.* **55** (1) 189–90.

Raw, F. (1960b). Earthworm population studies: a comparison of sampling methods. *Nature, Lond.* **187** (4733) 257.

Raw, F. (1961). The agricultural importance of the soil meso-fauna. *Soils & Fert.* **14,** 1–2.

Raw, F. (1962). Studies of earthworm populations in orchards. I. Leaf burial in apple orchards. *Ann. appl. Biol.* **50,** 389–404.

Raw, F. (1965). Current work on side effects of soil applied organophosphorus insecticides. *Ann. appl. Biol.* **55,** 342–3.

Raw, F. (1966). The soil fauna as a food source for moles. *J. Zool., Lond.* **149,** 50–4.

Raw, F. and Lofty, J. R. (1959). Earthworm populations in orchards. *Rep. Rothamsted exp. Stn. for 1958*, 134–5.

Reynoldson, T. B. (1955). Observations on the earthworms of North Wales. *North Wales Nat.* **3,** 291–304.

Reynoldson, T. B. (1966). The ecology of earthworms with special reference to North Wales habitats. *Rep. Welsh Soils Discuss. Grp.* 25–32.

Reynoldson, T. B., O'Connor, F. B. and Kelly, W. A. (1955). Observations on the earthworms of Bardsey. *Bardsey Obs. Rep.*, 9.

Rhee, J. A. van (1963). Earthworm activities and the breakdown of organic matter in agricultural soils. In *Soil Organisms*, J. Doeksen and J. van der Drift (eds). North Holland Publishing Co., Amsterdam, pp. 55–9.

Rhee, J. A. van (1965). Earthworm activity and plant growth in artificial cultures. *Pl. and Soil*, **22,** 45–8.

Rhee, J. A. van (1967). Development of earthworm populations in orchard soils. In *Progress in Soil Biology*, O. Graff and J. Satchell (eds). North-Holland Publishing Co., Amsterdam, pp. 360–71.

Rhee, J. A. van (1969). Inoculation of earthworms in a newly-drained polder. *Pedobiologia*, **9,** 128–32.

Rhee, J. A. van (1969). Development of earthworm populations in polder soils. *Pedobiologia*, **9,** 133–40.

Rhee, J. A. van (1972). Some aspects of the productivity of orchards in relation to earthworm activities. *Proc. 4th Int. Congr. Soil Zool.* (In press.)

Rhee, J. A. van and Nathans, S. (1961). Observations on earthworm populations in orchard soils. *Neth. J. agric. Sci.* **9** (2) 94–100.

Rhoades, W. C. (1963). A synecological study of the effects of the imported fire ant (*Solenopsis saevissima* Vichteri) eradication program. II. Light trap, soil sample, litter sample and sweep net methods of collecting. *Fla. Ent.* **46,** 301–10.

Ribaudcourt, E. and Combault, A. (1907). The role of earthworms in agriculture. *Bull. Soc. for. Belg.* 212–23.

Richards, J. G. (1955). Earthworms (recent research work). *N.Z. Jl. Agric.* **91,** 559.

Richardson, H. C. (1938). The nitrogen cycle in grassland soils: with special reference to the Rothamsted Park grass experiment. *J. agric. Sci., Camb.* **28,** 73–121.

Richter, G. (1953). The action of insecticides on soil macrofauna. *NachrBl. dt. PflSchutzdienst, Berl.* **7,** 61–72.

Robertson, J. D. (1936). The function of the calciferous glands of earthworms. *J. exp. Biol.* **13,** 279–97.

Robinson, J. S. (1953). Stimulus substitution and response learning in the earthworm. *J. Comp. Physiol. Psychol.* **46,** 262–6.

Rodale, R. (1948). Do chemical fertilizers kill earthworms? *Organic Gardening,* **12** (2) 12–17.

Rodale, R. (1961). *The challenge of earthworm research.* S. & H. Foundation, Penn. 102 pp.

Roots, B. I. (1955). The water relations of earthworms. I. The activity of the nephridiostome cilia of *L. terrestris* L. and *A. chlorotica* (Sav.) in relation to the concentration of the bathing medium. *J. exp. Biol.* **32,** 765–74.

Roots, B. I. (1956). The water relations of earthworms. II. Resistance to desiccation and immersion and behaviour when submerged and when allowed choice of environment. *J. exp. Biol.* **33,** 29–44.

Roots, B. I. (1957). Nature of chloragogen granules. *Nature, Lond.* **179,** 679–80.

Roots, B. I. (1960). Some observations on the chloragogenous tissue of earthworms. *Comp. Biochem. Physiol.* **1,** 218–26.

Ruschmann, G. (1953). Antibioses and symbioses of soil organisms and their significance in soil fertility. Earthworm symbioses and antibioses. *Z. Acker. PflBau.* **96,** 201–18.

Russell, E. J. (1910). The effect of earthworms on soil productiveness. *J. agric. Sci., Camb.* **2,** 245–57.

Russell, E. J. (1950). *Soil conditions and plant growth.* Longman, London, 8th edition.

Ryšavý, B. (1964). Some notes of the life history of the cestode *Dilepis undula* Shrank, 1788. *Helminthologia,* **5,** 173–6.

Ryšavý, B. (1969). Lumbricidae – an important parasitological factor in helminthoses of domestic and wild animals. *Pedobiologia,* **9** (1/2) 171–4.

Ryzhikov, K. M. (1949). Syngamidae of domestic and wild animals. *Moskva,* pp. 1–165.

Salisbury, E. J. (1925). The influence of earthworms on soil reaction and the stratification of undisturbed soils. *J. Linn. Soc.* (*Bot*), **46,** 415–25.

Saroja, K. (1959). Studies on oxygen consumption in tropical poikilotherms. 2. Oxygen consumption in relation to body size and temperature in the earthworm *Megascolex mauritii* when kept submerged in water. *Proc. Indian Acad. Sci.* B. **49,** 183–93.

Satchell, J. E. (1955). Some aspects of earthworm ecology. In *Soil Zoology,* D. K. Mc E. Kevan (ed.). Butterworths, London, pp. 180–201.

Satchell, J. E. (1955). *Allolobophora limicola.* An earthworm new to Britain. *Ann. Mag. nat. Hist.* (12) **8,** 224.

Satchell, J. E. (1955). The effects of BHC, DDT and parathion on the soil fauna. *Soils & Fert.* **18** (4) 279–85.

Satchell, J. E. (1955). An electrical method of sampling earthworm populations. In *Soil Zoology,* D. K. Mc E. Kevan (ed.). Butterworths, London, pp. 356–64.

Satchell, J. E. (1958). Earthworm biology and soil fertility. *Soils & Fert.* **21,** 209–19.

Satchell, J. E. (1960). Earthworms and soil fertility. *New Scientist* **7,** 79–81.

Satchell, J. E. (1963). Nitrogen turnover by a woodland population of *Lumbricus terrestris.* In *Soil Organisms,* J. Doeksen and J. van der Drift (eds). North Holland Publishing Co., Amsterdam, pp. 60–6.

Satchell, J. E. (1967). Lumbricidae. In *Soil Biology,* A. Burgess and F. Raw (eds). Academic Press, London and N.Y., pp. 259–322.

Satchell, J. E. (1969). Studies on methodical and taxonomical questions. *Pedobiologia,* **9,** 20–5.

Satchell, J. E. and Lowe, D. G. (1967). Selection of leaf litter by *Lumbricus terrestris.* In *Progress in Soil Biology,* O. Graff and J. E. Satchell (eds). North Holland Publishing Co., Amsterdam, pp. 102–19.

Saussey, M. (1957). A case of commensalism in the lumbricids. *Bull. Soc. ent. Fr.* **62** (1/2) 15–19.

Saussey, M. (1959). Observations sur les relations entre la composition physico-chimique du sol et son peuplement en Lumbricides. *Arch. Zool. exp. gen.* **93,** 123–34.

Saussey, M. (1966). Zoologie experimentale-relations entre la régéneration caudale et la diapause chez *Allolobophora icterica* (Savigny) (Oligochaete lombricien). *C.r. hebd. Séanc. Acad. Sci., Paris,* **263,** 1092–4.

Scharpenseel, H. W. and Gewehr, H. (1960). Studien zur Wasserbewegung im Boden mit Tritium-Wasser. *Z. PflErnähr. Düng.* **88,** 35–49.

Schmidt, H. (1955). Behaviour of two species of earthworm in the same maze. *Science,* **121,** 341–2.

Schmid, L. A. (1947). Induced neurosecretion in *Lumbricus terrestris.* *J. exp. Zool.* **104,** 365–77.

Schneider, K. C. (1908). *Histologisches Praktikum der Tiere.* Jena. 320 pp.

Schread, J. C. (1952). Habits and control of the oriental earthworm. *Bull. Connect. Agric. Exp. Stn.* **556,** 5–15.

Scott, H. E. (1960). Control of mites in earthworm beds. *North Carolina State Agr. Ext. Serv. Ext. Folder,* **181.**

Schwartz, B. and Alicata, J. E. (1931). Concerning the life history of lungworms of swine. *J. Parasit.* **18,** 21–7.

Scrickhande, J. C. and Pathak, A. N. (1951). A comparative study of the physico-chemical characters of the castings of different insects. *Indian J. agric. Sci.* **21,** 401–7.

Shindo, B. (1929). On the seasonal and depth distribution of some worms in soil. *J. Coll. Agric., Tokyo,* **10,** 159–71.

Shiraishi, K. (1954). On the chemotaxis of the earthworm to carbon dioxide. *Sci. Rep. Tôhoku Univ.* **20** (4) 356–61.

Sims, R. W. (1963). Oligochaeta (Earthworms) *Proc. S. Lond. ent. nat. Hist. Soc.* (2) 53.

Sims, R. W. (1963). A small collection of earthworms from Nepal. *J. Bombay nat. Hist. Soc.* **60** (1) 84–91.

Sims, R. W. (1964). Oligochaeta from Ascension Island and Sierra Leone including records of *Pheretima* and a new species of *Ichogaster*. *Ann. Mag. nat. Hist.* **7** (13) 107–13.

Sims, R. W. (1964). Internal fertilization and the functional relationship of the female and the spermathecal systems in new earthworms from Ghana (Eudrilidae: Oligochaeta). *Proc. zool. Soc. Lond.* **143** (4) 587–608.

Sims, R. W. (1966). The classification of the megascolecoid earthworms: an investigation of Oligochaete systematics by computer techniques. *Proc. Linn. Soc. Lond.* **177,** 125–41.

Sims, R. W. (1969). Outline of an application of computer techniques to the problem of the classification of the megascolecoid earthworms. *Pedobiologia,* **9** (5) 35–41.

Skarbilovic, T. S. (1950). The study on the biology of *Capillaria mucronata* and on the epizootology of capillarioses of the urinary bladder of sable and mink. *Trudy vsesoyuznogo inst. gelmintologii im akad. K.I. Skriabina,* **4,** 27–33.

Slater, C. S. (1954). Earthworms in relation to agriculture. *U.S.D.A. A.R.C. Circ.*

Slater, C. S. and Hopp, H. (1947). Leaf protection in winter to worms, U.S.A. *Proc. Soil Sci. Soc. Am.* **12,** 508–11.

Smallwood, W. M. (1923). The nerve net in the earthworm: preliminary report. *Proc. Nat. Acad. Sci. Washington,* p. 9.

Smallwood, W. M. (1926). The peripheral nervous system of the common earthworm, *Lumbricus terrestris*. *J. comp. Neurol.* **42,** 35–55.

Smith, F. (1915). Two new varieties of earthworms with a key to described species in Illinois. *Bull. Ill. State Lab. nat. Hist.* **10** (8) 551–9.

Smith, F. (1928). An account of changes in the earthworm fauna of Illinois. *Bull. Ill. State nat. Hist. Survey*, **17** (10) 347–62.

Smith, R. D. and Glasgow, L. L. (1965). Effects of heptachlor on wildlife in Louisiana. *Proc. 17th Ann. Conf. S/E Ass. Game and Fish Comm.* **17,** 140–54.

Stephenson, J. (1929). Oligochaeta: in reports of an expedition to Brazil and Paraguay, 1926–7. *J. Linn. Soc.* (*Zool*), **37,** 291–325.

Stephenson, J. (1930). *The Oligochaeta.* Oxford University Press. 978 pp.

Stephenson, J. (1945). Concentration regulation and volume control in *Lumbricus terrestris* L., *Nature, Lond.* **155,** 635.

Stickel, W. H., Mayne, D. W. and Stickel, L. F. (1965). Effects of heptachlor-contaminated earthworms on woodcocks. *J. Wildl. Manage.* **29,** 132–46.

Stockdill, S. M. J. (1959). Earthworms improve pasture growth. *N.Z. J. Agric.* **98,** 227–33.

Stockdill, S. M. J. and Cossens, G. G. (1966). The role of earthworms in pasture production and moisture conservation. *Proc. N.Z. Grassl. Ass.* 168–83.

Stockdill, S. M. J. (1966). The effect of earthworms on pastures. *Proc. N.Z. ecol. Soc.* **13,** 68–74.

Stöckli, A. (1928). Studien über den Einfluss der Regenwürmer auf die Beschaffenheit des Bodens. *Landw. Jb. Schweiz.* **42** (1).

Stöckli, A. (1949). Einfluss der Mikroflora und Fauna auf die Beschaffenheit des Bodens. *Z. PflErnähr. Düng.* **45** (90) 41–53.

Stöckli, A. (1958). Die Regenwurmarten in landwirtschaftlich genutzten Böden des schweizerischen Mittellandes. *Separatabdruck aus dem Landwirtschaftlichen Jahrbuch der Schweiz.* **72** (7) 699–725.

Stokes, B. M. (1958). The worm-eating slugs *Testacella scutulum* Sowerby and *T. haliotidea* Drapernaud in captivity. *Proc. malac. Soc. Lond.* **33** (1) 11–20.

Stolte, H. A. (1962). Oligochaeta. In *Bronn's Klassen und Ordnungen des Tierreichs,* **4** (3) 891–1141. Geest and Portig, Leipzig.

Støp-Bowitz, C. (1969). Did Lumbricids survive the quarternary glaciations in Norway. *Pedobiologia,* **9,** 93–8.

Stough, H. B. (1926). Giant nerve fibres of the earthworm. *J. comp. Neurol.* **40.**

Stringer, A. and Pickard, J. A. (1963). The DDT content of soil and earthworms in an apple orchard at Long Ashton. *Long Ashton Res. Sta. Rep.* 127–31.

Sun, K. H. and Pratt, K. C. (1931). Do earthworms grow by adding segments? *Am. Nat.* **65,** 31–48.

Svendsen, J. A. (1955). Earthworm population studies: a comparison of sampling methods. *Nature, Lond.* **175,** 864.

Svendsen, J. A. (1957). The distribution of Lumbricidae in an area of Pennine Moorland (Moor House, Nature Reserve). *J. Anim. Ecol.* **26** (2) 409.

Svendsen, J. A. (1957b). The behaviour of lumbricids under moorland conditions. *J. Anim. Ecol.* **26** (2) 423–39.

Swaby, R. J. (1949). The influence of earthworms on soil aggregation. *J. Soil Sci.* **1** (2) 195–7.

Swartz, R. D. (1929). Modification of the behaviour of earthworms. *J. Comp. Psychol.* **9,** 17–33.

Takano, S. and Nakamura, Y. (1968). A new host earthworm, *Allolobophora japonica* Michaelsen, (Oligochaeta: Lumbricidae), of the calypterate muscoid fly, *Onesia subalpina* Kurahashi. (Diptera: Calliphoridae.) *Appl. ent. Zool.* **3** (1) 51–2.

Tembe, V. B. and Dubash, P. J. (1961). The earthworms: a review. *J. Bombay nat. Hist. Soc.* **58** (1) 171–201.

Tenney, F. G. and Waksman, S. A. (1929). Composition of natural organic materials and their decomposition in the soil. IV. The nature and rapidity of decomposition of the various organic complexes in different plant materials, under aerobic conditions. *Soil Sci.* **28,** 55–84.

Teotia, S. P., Duley, F. L. and McCalla, T. M. (1950). Effect of stubble mulching on number and activity of earthworms. *Neb. agric. Exp. Sta. Res. Bull.* **165,** 20.

Thompson, A. R. (1971). Effects of nine insecticides on the numbers and biomass of earthworms in pasture. *Bull. Env. Cont. Toxicol.* **5** (6) 577–86.

Tischler, W. (1955). Effect of agricultural practice on the soil fauna. In *Soil Zoology*, D. K. Mc. E. Kevan (ed.). Butterworths, London, 125–137.

Tracey, M. V. (1951). Cellulase and chitinase in worms. *Nature, Lond.* **167,** 776–7.

Trifonov, D. (1957). Über die Bekämpfung der Maulwurfsgrille und des Regenwurms mit dem Präparat. *Alon Kombi Bulgar. Tiûtiûn*, **2,** 114–15.

Tromba, F. G. (1955). Role of the earthworm *Eisenia foetida*, in the transmission of *Stephanurus dentatus*. *J. Parasit.* **41,** 157–61.

Uhlen, G. (1953). Preliminary experiments with earthworms. *Landbr. Høgsk. Inst. Jordkultur Meld.* **37,** 161–83.

Urquhart, A. T. (1887). On the work of earthworms in New Zealand. *Trans. N.Z. Inst.* **19,** 119–23.

Villot, F. C. A. (1883). Memoire sur les cystiques des ténias. *Ann. Sci. Nat. Zool.* **15,** 1–61.

Volz, P. (1962). Contributions to a pedo-zoological study of sites based on observations in the south-eastern Palatinate. *Pedobiologia*, **1,** 242–90.

Waite, R. H. (1920). Earthworms. The important factor in the transmission of gapes in chickens. *Maryland agric. exp. Bull.* **234,** 103–18.

Waksman, S. A. and Martin, J. P. (1939). The conservation of the soil. *Science*, **90,** 304–5.

Walton, W. R. (1928). Earthworms as pests and otherwise. *U.S.D.A. Farmers' Bulletin 1569, Washington, D.C.*, 14.

Walton, W. R. (1933). The reaction of earthworms to alternating currents of electricity in the soil. *Proc. ent. Soc. Wash.* **35,** 24–7.

Waters, R. A. S. (1952). Earthworms and the fertility of pasture. *Proc. N.Z. Grassl. Ass.* 168–75.

Waters, R. A. S. (1955). Numbers and weights of earthworms under a highly productive pasture. *N.Z. J. Sci. Technol.* **36** (5) 516–25.

Watkin, B. R. (1954). The animal factor and levels of nitrogen. *J. Br. Grassld Soc.* **9,** 35–46.

Way, M. J. and Scopes, N. E. A. (1968). Studies on the persistence and effects on soil fauna of some soil applied systemic insecticides. *Ann. appl. Biol.* **62,** 199–214.

Weber, G. (1953). The macrofauna of light and heavy arable soils and the effect on them of plant protection substances. *Z. PflErnähr. Düng.* **61,** 107–18.

Weisbach, W. W. (1962). Regenwürmer und Essbare Erde. *Biol. Jaarb. Dodonea.* **30,** 225–38.

Went, J. C. (1963). Influence of earthworms on the number of bacteria in the soil. In *Soil Organisms*, J. Doeksen and J. van der Drift (eds). North Holland Publ. Co., Amsterdam, pp. 260–5.

Wheatley, G. A. and Hardman, J. A. (1968). Organochlorine insecticide residues in earthworms from arable soils. *J. Sci. Fd. Agric.* **19,** 219–25.

Wherry, E. T. (1924). Soil acidity preferences of earthworms. *Ecology*, **5,** 89–90.

Whitney, W. K. (1967). Laboratory tests with Dursban and other insecticides in soil. *J. econ. Ent.* **60,** 68–74.

Wilcke, D. E. von (1952). On the domestication of the 'soilution' earthworm. *Anz. Schädlingsk.* **25,** 107–9 (G).

Wilcke, D. E. von (1953). Zur Kenntnis der Lumbricidenfauna Deutschlands. *Zool. Anz.* **151,** 104–6.

Wilcke, D. E. von (1955). Critical observations and proposals on the quantative analysis of earthworms populations in soil zoology studies. *Z. PflErnähr. Düng.* **68,** 44–9 (G).

Witkamp, M. (1966). Decomposition of leaf litter in relation to environment, microflora and microbial respiration. *Ecology*, **47**, 194–201.

Wittich, W. (1953). Untersuchungen über den Verlauf der Streuzersetzung auf einem Boden mit Regenwurmtätigkeit. *Schrift Reihe forstl. Fak. Univ. Gottingen*, **9**, 7–33.

Wojewodin, A. W. (1958). Ungefährlichkeit der Herbizide für die Biozönose. *Int. Konf. Herb.* 97–102.

Wolf. A. V. (1937). Notes on the effect of heat on *L. terrestris*. *Ecology*, **19**, 346–8.

Wolf, A. V. (1938). Studies on the behaviour of *L. terrestris* to dehydration; and evidence for a dehydration tropism. *Ecology*, **19**, 233–42.

Wolf, A. V. (1940). Paths of water exchange in the earthworm. *Physiol. Zool.* **13**, 294–308.

Wolf, A. V. (1941). Survival time of the earthworm as affected by raised temperatures. *J. cell. comp. Physiol.* **18**, 275–8.

Wollny, E. (1890). Untersuchungen über die Beeinflussung der Fruchtbarkeit der Ackerkrume durch die Tätigkeit der Regenwürmer. *Forsch. Agrik. Physik.* **13**, 381–95.

Woodhead, A. A. (1950). Life history cycle of the giant kidney worm, *Dioctophyma renale* (Nematoda) of man and many other animals. *Trans. Am. microsc. Soc.* **69**, 21–46.

Yerkes, R. M. (1912). The intelligence of earthworms. *J. Anim. Behav.* **2**, 332–52.

Zhinkin, L. (1936). The influence of the nervous system on the regeneration of *Rhynchelmis limosella*. *J. exp. Zool.* **73**, 43–65.

Zicsi, A. (1954). The role of earthworms in the soil, as investigated by soil analyses, experiments and survey at the University in Gödöllö. *Agrartud. Egypt. agron. Kar Kiadv.* **1** (14) 1–20.

Zicsi, A. (1962). Determination of number and size of sampling unit for estimating lumbricid populations of arable soils. In *Progress in Soil Zoology*, P. W. Murphy (ed.). Butterworths, London, 68–71.

Zicsi, A. (1958). Einfluss der Trockenheit und der Bodenarbeitung auf das Leben der Regenwürmer in Ackerboden. *Acta Agron.* **7**, 67–74.

Zicsi, A. (1962). Über die Dominanzhältnisse einheimischer Lumbriciden auf Ackerboden. *Opusc. Zool. Budapest.* **4** (2–4) 157–61.

Zicsi, A. (1969). Über die Auswirkung der Nachfrucht und Bodenbearbeitung auf die Aktivität der Regenwürmer. *Pedobiologia*, **9** (1–2) 141–6.

Zrazhevskii, A. I. (1957). *Dozhdevye chervi kak faktor plodorodiya lesnykh pochv*. Kiev. 135 pp.

Indexes

Systematic index

(*Italic page numbers indicate the more important references in the text whereas bold numbers refer to illustrations.*)

Author index

General index

(Italic page numbers indicate the more important references in the text whereas bold numbers refer to illustrations.)